T0251411

REDUCE for Physicists

REDUCE for Physicists

N MacDonald

Department of Physics and Astronomy
University of Glasgow

CRC Press
Taylor & Francis Group
Boca Raton   London   New York

CRC Press is an imprint of the
Taylor & Francis Group, an **informa** business

CRC Press
Taylor & Francis Group
6000 Broken Sound Parkway NW, Suite 300
Boca Raton, FL 33487-2742

© 1994 by Taylor & Francis Group, LLC
CRC Press is an imprint of Taylor & Francis Group, an Informa business

No claim to original U.S. Government works

Visit the Taylor & Francis Web site at
http://www.taylorandfrancis.com

and the CRC Press Web site at
http://www.crcpress.com

Contents

Preface

In recent years many research workers in physics have successfully employed symbolic computation. Languages such as REDUCE, MAPLE and MACSYMA have become widely available in versions for mainframe computers, work stations, and more recently for microcomputers. These languages have vast power to remove tedium and casual error from extensive mathematical manipulation. It seems timely to produce a text aimed at a readership who approach the topic of symbolic computation as a convenient tool for tackling physics problems. The REDUCE language was first developed, by A C Hearn, in a particular context in theoretical physics, quantum electrodynamics, and it has had many useful applications in physics and engineering. Cohen and Fitch (1991) give a comprehensive survey of applications of REDUCE and other systems, in the years 1979–1988. About half the papers they list involve the use of REDUCE. The topics range very widely, and include general relativity, quantum electrodynamics, fluids and plasmas, optics, mechanics, classical and quantum perturbation theory.

This text assumes that the reader has access to some edition of the REDUCE language on a suitable computer, and to the manual appropriate to that edition. In developing the programs the version used is REDUCE 3.3, as supplied at the end of 1988. However the programs will run in the version 3.4, supplied in June 1991. Significant changes from 3.3 to 3.4 are mentioned in notes at the end of certain chapters, with some additional comments on 3.4 in an Appendix.

The basic premise of this text is that while the manual defines the content of REDUCE, the reader seeks assistance in preparing to use the language for tackling real theoretical problems. The reader is assumed to have some experience in solving problems in physics, so that the text draws on applications in physics without devoting an undue amount of space to setting out the physical context. Each chapter introduces a few aspects of the language and illustrates them with a small number of applications.

Chapter 1 introduces hands-on and programmed working, giving an overall impression of what it is like to work with REDUCE. The illustrative examples are drawn from Hamiltonian mechanics.

Chapter 2 surveys a number of basic topics to do with arithmetic, elementary transcendental functions and integration.

Chapter 3 is mainly concerned with a facility for solving algebraic equations, which in turn makes it necessary to describe how REDUCE handles lists. The physical application is to dimensional analysis.

Chapter 4 introduces repetitive processes. These are processes that run through a range of values of an index—loops—and processes which are repeatedly summoned by a main program—procedures. The illustrative examples are from Lagrangian and Hamiltonian dynamics. Since one of these examples uses a rotation matrix, the matrix facilities of REDUCE are introduced at this stage.

Chapter 5 continues with matrices, employing determinants and characteristic polynomials in a linear stability investigation of a generalised form of the Lorenz nonlinear model. In the course of preparing the parameters of a linearised dynamical system for this investigation one has to filter out unsuitable cases, which leads to a discussion of conditional statements.

Chapter 6 turns to quantum mechanics, with the need for noncommuting operators drawing attention to the process of defining operators in REDUCE. The illustrations include the quantum analogue of a dynamics problem from chapter 4, and a perturbation calculation.

Chapter 7 deals with a method for obtaining approximate solutions to nonlinear ordinary differential equations, the method of harmonic balance. While this method has a long history, it is of current interest in a variety of physical and engineering applications. The programs here require more techniques for manipulating polynomials, and also introduce arrays and parts.

Chapter 8 examines first how to translate REDUCE output into FORTRAN, and then the use of the GENTRAN package to write programs in FORTRAN that use REDUCE to evaluate expressions.

In most of these chapters there are exercises, which may be short self-contained topics, or may require copying and amending one of the illustrative programs. On the whole, if the comments are left out, it should not take too long to type in a selection of the programs and begin to develop one's own extensions and modifications. The subject of chapter 7 is one with which I am currently working, and affords an opportunity for more extended non-trivial exercises, which could be suitable for undergraduate computational projects.

Each chapter ends with a summary of the instructions and conventions that are introduced in the chapter. In chapters 2 and 4 this summary is followed by a brief commentary on the more important features that differ in REDUCE 3.4.

The appendices give some further comments on the new facilities available with REDUCE 3.4, and a more substantial application of REDUCE, to a problem in plasma wave theory, in the form of a paper by Declan Diver.

Some topics are deliberately excluded from this text, which is concerned solely with the use of one existing language and not with general aspects of symbolic computation, nor with the extension of the facilities of the language. For an introduction to general aspects of symbolic computation, see Davenport et al (1988). No attempt is made here to assess the relative merits of REDUCE and competing languages. A handy consumer's guide to the available languages is available in Harper, Wooff and Hodgkinson (1991). We hope that teachers using other languages may find some of our topics stimulating as a source of exercises and projects.

While learning to handle REDUCE, and in the preparation of some parts of this text, I benefited greatly from the experience and enthusiasm of a colleague, Declan Diver. He has also kindly allowed me to reproduce an extensive program of his, for wave propagation in a non-uniform plasma. I have incorporated two shorter programs from the published literature (programs 4.2, 6.3) which are acknowledged below. I wish to thank John Anderson, of the Aerospace Engineering Department, University of Glasgow, for bringing the wind tower application to my attention. I have adapted and developed further a number of programs written by Seonaid McLeod, Kenneth Skeldon and Alan Wilson in undergraduate projects in this department (programs 3.1, 4.3, 6.1. 6.2. and 7.2). I wish to thank these students, and two research students, Graeme Stewart and David Ramsay, for helpful comments. I also wish to thank several people unknown to me, who have read versions of the typescript for the publisher, and who have provided a great variety of helpful comments and suggestions.

Acknowledgments

Program 4.2 is taken from Garrad and Quarton (1986), and program 6.3 from Sage (1988). These and Diver (1991), which is given as Appendix 3, are reproduced by kind permission of Academic Press Ltd.

1

A first look

In which we examine what it is like to work with REDUCE, and introduce hands-on and programmed calculations.

1.1 HANDS-ON CALCULATION, ASSIGNMENTS AND VALUES

REDUCE is a language for symbolic computation, which enables a computer to perform symbolic operations of algebra and calculus. Its repertoire includes, for example, the following operations:

differentiate a function
integrate a function
find the determinant of a matrix
solve a set of linear algebraic equations.

In REDUCE the instruction df(x**2,x), standing for dx^2/dx, yields the result $2x$. On the other hand a conventional (arithmetical) computation employs a numerical algorithm to obtain an approximation to the numerical value of the derivative at a specific value of x. It steps forward through a range of values to build up an approximation to the derivative over that range.

REDUCE allows alternative ways to go about a calculation, interactive or hands-on calculation on the one hand, and programmed calculation on the other. Fortunately one can become familiar with these two modes of calculation very quickly, using only a few of the facilities available. In either mode the first move is to enter

reduce

to which the response is a line identifying the edition used, for example

REDUCE 3.3 15-Jan-88

followed by a cue number

1:

The cue number signifies that the computer will accept a command. In the hands-on mode, each successive command yields, so long as it is in a correct form, either an output line and a new cue number, or just a new cue number. Here is a short sequence of commands with their output.

```
1: a:= 1 + x*2 + x**2;
          2
   A := X   + 2*X + 1
2: on factor;
3: a:= a;
           2
   A := (X + 1)
4: b:= df(a,x);
   B := 2*(X + 1)
5: c:= df(a,x,2);
   C := 2
6: x:= 1;
   X := 1
7: a:= a;
   A := 4
8: clear a;
9: a:= a;
   A := A
10: b:= b;
    B := 4
11:
```

Each command lies between a cue number (n:) and a terminator (;). In this example there are four assignment commands a:= ..., b:= ..., c:= ..., x:= ..., each of which yields an output. These commands assign to the variable on the left the value of the expression on the right or in the case of x:= 1; the numerical value on the right. Until assigned a value, a variable is its own value, and is said to be clear. Thus after the first command in this sequence a has the value $x^2 + 2x + 1$, while x is clear, having the value x. After the sixth command, x has the value 1 and consequently a has the value 4. The fifth command assigns to c a value, the second x-derivative of the current value of a, that is of $x^2 + 2x + 1$, and so c acquires a numerical value 2. Any variable can be returned to its original value by the command clear. This may have side effects, but in 10: we see that consequences of the previously assigned value are not wiped out.

To state that x has the value x may seem pedantic or even perverse, but it is important in keeping track of the effects of a sequence of commands. We shall illustrate this in section 1.6 where we compare the assignment command x:= y; with the command let x = y;

The input can mix lower case and capitals, but one should note that, for example, a and A are not accepted to denote two different quantities. Throughout this book, to keep clear the distinction between input and output lines, we shall use lower case in the input, except for occasional symbols, such as a counting index J. The input is in conventional single-line form as used in other computing languages, with * indicating product, ** or ^ indicating power, / indicating ratio. When an expression a is to be divided by the product of expressions b and c, use a/(b*c) — the bracket is essential.

All literal symbols are in capitals in the output. The output resembles conventional mathematical notation in using a raised power, not ** or ^, but still uses * for product. According to the length of the expressions in numerator and denominator, a ratio may appear in the output as A/B or as $\frac{A}{B}$. Terms in an output expression are quite likely not to appear in the order used in the input expression.

In our example there is a switch command, on factor, which alters the way in which the output is displayed, so that when a is recalled, by a:= a, it appears in factored form. All subsequent factorable expressions will appear in factored form, until one enters the command off factor. A switch command has no output. We shall later encounter a variety of these switch commands, which are sometimes called flags. In the sequence above we recall a, for example, by the command

```
7: a:= a;
```

which gives output

    A := 4

We can instead simply type

a;

which gives the output

    4

This is fine as we go through a simple hands-on calculation, but if we read through a lengthy output we find it helpful to have the more complete form.

In the hands-on mode we may not wish to see all output lines. We can use the terminator $ in place of ; to suppress the output from any command.

As an example of a situation in which the hands-on mode only requires two input cues, suppose that we encounter an integrand expressed in terms of elementary functions, and the integral does not spring to mind. For example the integrand might be

$$(z^2 + 1) \exp(-z) \cos(3z).$$

Instead of searching for a reference book, we can turn on REDUCE and carry out the sequence

```
1: x:= (z**2 + 1)exp(-z)cos(3*z)$
2: y:= int(x,z);
                  2
   Y:= -(25*COS(3*Z)*Z - 40*COS(3*Z)*Z + 12*COS(3*Z)
                  2
       - 75*SIN(3*Z)*Z  - 30*SIN(3*Z)*Z - 66*SIN(3*Z) )
             Z
     / (250*E  )
```

indicating that in this context REDUCE identifies the three transcendental functions cosine, sine and exponential, and can perform this integration. We shall say more on this topic in the next chapter.

1.2 LOOKING BACK AND MAKING MISTAKES

The screen can only display a limited number of lines, especially as the output is given generous space, because of the way powers and ratios are displayed. The ratio of two short expressions containing powers needs one line as input but five lines as output. We have to keep track of what happens at each step if we wish to refer back. We refer back by the obvious means of using symbols already used, or by using the notation ws(n). The abbreviation ws means workspace, while n is the cue number. In our example, ws(1) is the right hand side of the output line which follows the cue 1, that is

```
 2
X  + 2*X + 1
```

if we call for it immediately after the first assignment command. Similarly ws(3) is

```
       2
(X + 1)
```

However if we call for ws(1) after the factor switch is turned on, it is modified by the effect of that switch, and so is identical with ws(3). The command d:= ws(5)/ws(4) assigns to d the same expression as the command d:= c/b.

If we enter several commands together, properly separated by terminators, the output appears as separate numbered lines. For example if we start the sequence above with

```
1: a:= 1 + x*2 + x**2; on factor; a:= a;
```

the output appears as

```
        2
   A := X  + 2*X + 1
2:
3:
            2
   A := (X + 1)
```

so that ws(1) and ws(3) have the same significance as before. If we shorten the command 3: thus

```
3: a;
```

giving the output

$$(X + 1)^2$$

then ws(3) is the same as before.

Suppose that after a cue number we decide to terminate the hands-on session, either because we have found the required result, or because we have made some mistake and have to rethink our calculation. Then we must type

bye;

The response to this is

Bye

followed by the normal system prompt—the computer has now left REDUCE.

We should now consider what can go wrong in a hands-on session, and how to escape from trouble. It may seem pessimistic to bring up this topic so soon, but we want the reader to get his hands on the keyboard as quickly as possible, and at this stage elementary errors can be most disconcerting.

An obvious trivial error, especially for anyone used to a language without terminators, is to omit the terminator. Then nothing happens, except that the prompt moves down a line. All that is needed to correct this is to enter the terminator. So for example we could have, in the sequence in section 1.1,

```
5:c:= df(a,x,2)
;
    C := 2
6:
```

proceeding safely to the correct output. Another obvious trivial error is to misspell. For example, if you have a variable called square, and you type

squar;

the output is

```
SQUAR
```

This means that the new unintended variable squar is clear, and so its name is printed as its value.
Any illegitimate instruction, for example

```
5: c:= df(a,2);
```

in the sequence in section 1.1, ought to give an error message, in this case

```
*** 2 invalid as kernel
```

and a new cue number. This particular error message means that the integer 2 appears where a variable (in this context x) should appear. In the same sequence, asking for ws(2)/ws(5) yields the error message

```
*** Entry 2 not found
```

and a new cue number. There are some loopholes in the error detecting system. The mistaken instruction

```
2: b:= df(a);
```

yields, not an error message, but a mistaken output

```
            2
B := (X + 1)
```

As another example of an error message, suppose that you omit a bracket, entering

```
ws(20;
```

This will yield the message

```
WS(20$$$;
**** Too few right parentheses
```

As you gain experience with REDUCE, you will encounter a variety of error messages. Normally the context, or reference to the manual, will indicate how the preceding commands need to be modified. It is possible to enter an instruction which does not elicit an error message, but which causes REDUCE to go into an apparently interminable, or at any rate an unacceptably long, spell without output. Then interrupting, for example if you are working in a UNIX environment striking control + c, will give a message advising entering :H for help. The only obviously useful information thus obtained is to enter :q to quit; this should normally give a new cue number. It is also possible for an error to elicit a cryptic error message in LISP, which is the underlying language in terms of which REDUCE is constructed, and that :q will not rescue the session. As a last resort one can log out, log in and start over, thus losing the hands-on sequence.

1.3 A HANDS-ON CALCULATION; CONSTANT OF MOTION

We are quite likely to require a fairly long sequence of hands-on commands, when checking a 'one-off' unsupported statement in a textbook or research paper. For example, in Chang *et al* (1982) there is a comment that J M Greene communicated to the authors the expression

$$G = x^4 + 4x^2y^2 - 4p_x(p_xy - p_yx) + 4x^2y + 3(p_x^2 + x^2) \qquad (1.1)$$

as a first integral (constant of motion) for the dynamical system with Hamiltonian

$$H = (p_x^2 + p_y^2 + x^2 + y^2)/2 + Dx^2y - Cy^3/3 \qquad (1.2)$$

in the case $D = 1$, $C = -6$. For readers not familiar with the Hamiltonian form of dynamics, in this example the Hamiltonian is the total energy. The part expressed in terms of momenta p_x, p_y is the kinetic energy and the rest the potential energy. The mass is taken as one kg. When $D = C = 0$ this is the energy of a linear oscillator. Hamilton's equations, which we shall use shortly, are

$$\begin{aligned} \mathrm{d}x/\mathrm{d}t &= \partial H/\partial p_x & \mathrm{d}p_x/\mathrm{d}t &= -\partial H/\partial x \\ \mathrm{d}y/\mathrm{d}t &= \partial H/\partial p_y & \mathrm{d}p_y/\mathrm{d}t &= -\partial H/\partial y. \end{aligned} \qquad (1.3)$$

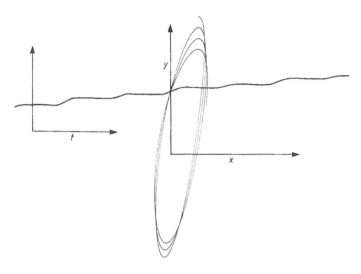

Figure 1.1 A schematic example of variable $y(t)$ plotted against variable $x(t)$ for the system of Henon and Heiles (1964), with D and C not taking the special values 1, -6 (at the special values, a closed loop is obtained). Figure 1 also shows G plotted against time over the same interval, for the same D and C.

In this example they consist of a pair of equations defining the momenta and a pair giving accelerations in terms of the gradient of the potential. This second pair of equations are equivalent to the Lagrangian equations or the Newtonian equations

$$m\mathrm{d}^2x/\mathrm{d}t^2 = F.$$

In our example the Hamiltonian equations are

$$\begin{aligned}
\mathrm{d}x/\mathrm{d}t &= p_x & \mathrm{d}p_x/\mathrm{d}t &= -x - 2Dxy \\
\mathrm{d}y/\mathrm{d}t &= p_y & \mathrm{d}p_y/\mathrm{d}t &= -Dx^2 + Cy^2 - y.
\end{aligned} \tag{1.4}$$

Motion is bounded if the initial values lie inside the triangle bounded by the lines

$$y^2 = Dx^2/C \qquad y = 1/2D.$$

The main content of the paper by Chang *et al* is to identify a very small set of values of the parameters C and D for which this dynamical system, the system of Henon and Heiles (1964), has a first integral other than the energy. The paper does not provide an algorithm for constructing the first integral when one does exist. This system has been intensively studied as a prototype for conservative dynamical systems with highly complex behaviour. Such behaviour is likely to occur when there is no first integral other than the energy, so that a search for such a first integral is of more than formal interest. In figure 1.1, we show one case of the complex behaviour of the bounded x, y variables when there is no first integral, along with the time dependence of G in that case. A closed loop is obtained in the $x - y$ plot for the special D and C values for which there is a first integral, and G becomes constant.

To check that $dG/dt = 0$, we work through the following hands-on sequence.

```
1: h:= (x**2 + y**2 + px**2 + py**2)/2 +d*x**2*y -c*y**3/3;
             3       2      2     2   2     2
   H := - 2*C*Y - 6*D*X *Y -3*PX - 3*PY -3*X - 3*Y
       ----------------------------------------
                     6
2: g:= x**4+4*x**2*y**2-4*px*(px*y-py*x)+4*x**2*y+3*(px**2+
       x**2);
            2       2           4   2 2      2
   G := -(4*PX *Y -3*PX - 4*PX*PY*X -X -4*X *Y - 4*X *Y
         2
        -3*X )
3: df(x,t):= df(h,px);
   DF(X,T) := PX
4: df(y,t):= df(h,py);
   DF(Y,T) := PY
5: df(px,t):= -df(h,x);
   DF(PX,T) := -X*(2*D*Y + 1)
6: df(py,t):= -df(h,y);
                   2     2
   DF(PY,T) := C*Y - D*X - Y
7: D:= 1;
   D := 1
```

```
8: C:= -6;
   C := - 6
9: gg:= df(g,t);
   GG := 0
10: bye;
   Bye.
```

Here the momenta are denoted by px, py, since we cannot use subscripts in the REDUCE input. We shall see in a later chapter that the alternative notation with $x(1)$ and $p(1)$ in place of x, p_x, is available in REDUCE. At times we cannot avoid referring in the text both to p_x and to px, because we are discussing both the theory and its implementation in REDUCE. Steps 1: and 2: input the Hamiltonian H and the assumed constant function G, by assigning to variables h, g, values which are expressions in terms of x, y, px, py. Steps 3: to 6: are Hamilton's equations. In each of these df is used in two distinct ways. On the right hand side, because h is an explicit function of x, y, px and py, the appropriate partial derivatives are obtained. On the left hand side of 3: for example, df(x,t) implies that the variable x is to be treated as a function of an independent variable t, and that df(x,t) stands for the derivative of x with respect to t. The := in 3: means that this derivative is assigned the value of the partial derivative on the right. These four commands specify x, y, p_x and p_y as solutions of a set of first order coupled ordinary differential equations. Steps 7: and 8: assign integer values to D and C, and step 9: asks for the time derivative of G. Since dG/dt is a function of the four dependent variables and their derivatives with respect to t, and since explicit forms are now known for these time derivatives in terms of the dependent variables, the answer is computed. All cancellations are automatically performed, and the outcome GG:= 0 confirms the expected result. If in 2: we had used some other expression on the right, the outcome would be to assign to GG a non-zero value, which would be an explicit function of x, y, px, py.

When we discuss short sequences of commands in the forthcoming chapters, we shall normally drop the cue numbers.

1.4   A PROGRAMMED CALCULATION; CONSTANT OF MOTION

Obviously it requires careful typing to get the sequence of commands correct, in our last example. Suppose that we expect to check a variety of 'candidates' for constants of motion, for several Hamiltonians which, like the one discussed above, have two degrees of freedom. There are obvious advantages in recording commands 3: and onwards, to be reused with different input at steps 1: and 2: . Suppose we give the name contest to a file, previously prepared, containing a program made up of the next seven commands. Then we can enter steps 1: and 2: each time hands-on, but at cue 3: we give the command

```
in contest;
```

Within the file contest all commands are given without cue numbers. In this example we also suppress their output, except for the final command. The output of 3: is thus the output of that command, either GG:= 0 if the candidate passes the test, or GG:= some expression, if it does not. We conclude with

```
4: bye;
```

This last step is advisable even if we wish to go ahead immediately to run another program. This is especially vital if the second is a related program with the same notation, but not specifically intended to follow the first program. During a single REDUCE run, from 1: to bye;, no commands are forgotten. If not explicitly cancelled, for example by the command clear, they apply beyond the end of the particular program in which they occur. It is safest to terminate the REDUCE run and start up a new one.

The program contained in the file contest takes the following form, where we illustrate two conventions for inserting comments.

```
comment    a program to test a given function G as a possible
constant of motion for a system with Hamiltonian H, for
specific values of two parameters C and D$
df(x,t):= df(h,px)$    % four Hamilton equations
```

```
df(y,t):= df(h,py)$
df(px,t):= -df(h,px)$
df(py,t):= -df(h,py)$
D:= 1$ C:= -6$  %  prescribed values of two parameters needed
gg:= df(g,t);  %  should be zero
end;
```

The following general points are to be noted. Every program must end with end; . This is not the same as ending with bye;, for we are still in REDUCE, as indicated by the appearance of another cue number. We can continue with a hands-on sequence or input another file. It is vital to check all terminators, since the program will not stop, like a hands-on calculation, but will run two successive commands into one, and the error may only be detected in some indirect way. A comment does not need a terminator if it appears after %, but does if it appears after comment. Omitting this terminator can cause confusion, since REDUCE will attempt to combine the comment with the next command. This usually means, not that an error message appears, but that the next command is ignored. For this reason, the use of % is recommended, and comment should be regarded as obsolete. However note that a % must appear at the beginning of each line in a long comment.

Provided we take care of terminators, we can print more than one command on one line, making the file more compact. (In fact if we accidentally type a double terminator ; ; the second ; will be treated as a dummy command and the program will continue.) We can get (by using the terminator ;) intermediate outputs if we wish. Many programs will need more than one output.

With a longer program, or one giving an elaborate output, we may well wish to keep a record of the output on another file. To do this we use the sequence

```
1: out hamlet;
2:h:= (x**2 + y**2 + px**2 + py**2)/2 +d*x**2*y - c*y**3/3;
3:g:= x**4 + 4*x**2*y**2 - 4*px*(px*y-py*x) + 4*x**2*y
+3*(px**2+x**2);
4: in contest;
5: bye;
    Bye.
```

No output appears on the screen, but the new file hamlet contains a complete record of all input and all output, each line terminated by ; being followed by its output. In discussing complete programs we shall frequently present them in this form. This input/output file takes up much more space than the file containing the input program alone. The output lines are displayed in a very open manner, as we have seen. Also any commands typed on one line in the input program are separated in the output file, even if their output is suppressed.)

EXERCISE 1.1 Although we want to keep the specification of *H* and *G* separate from the main program, we may still want to reuse this preliminary input several times, with new versions of the main program. We can write a small program hgin, only containing these two inputs, and load it first, followed by contest. Do this. Note that here the two programs hgin and contest are designed to run in sequence, and so can safely be run in one REDUCE session.

When developing a program it is useful to be able to work in a manner intermediate between hands-on and programmed modes. To do this insert in the program, at suitable points, the command pause. This has the effect of eliciting a choice of continuing, by entering y, or of stopping the calculation, by entering n. In the first case the program continues to the next pause, to the end, or to an error. In the second case we get a new cue number. We have left the program and are in hands-on mode. To return to the program, after any hands-on commands have been performed, use the command cont. This command is only to be used after pause. Normally when running a program that includes pauses we send the output to the screen. Instead of inserting the pauses individually we can use the command

on demo;

at a suitable point in the program. From this point on there is a pause at every step, until we come to the command

off demo;

If we send the output to a file, and have not removed all pause commands, at any occurrence of pause the question

y or n?

will show on the screen, and typing y will cause the program to continue, while n will result in a cue number. Nothing appears on the screen to indicate the step at which the decision is to be made.

1.5 CHANGING THE FORM OF THE OUTPUT

As an example of an input program and its output recorded together in an output file, we present program 1.1, which is a simple extension of the program contest. Here we check whether both $D = 1$ and $C = -6$ are strictly necessary to give $dG/dt = 0$, rather than, for example, $C = -6D$ or $C + D = -5$. Thus we need to see GG before setting the two parameters, and if possible see it in a form that makes clear that certain terms will vanish for specific values of D and C. REDUCE has a variety of options for altering the manner in which an expression is organised; we have already seen the switch factor in use. The appropriate command here is

```
factor x,y,px,py;
```

which brings together all terms which have the same product of powers of the variables x, y, p_x, p_y, differing only in the constant factors.

PROGRAM 1.1

```
% examines why D = 1, C = - 6 make dg/dt = 0
h:= (x**2 + y**2 + px**2 + py**2)/2 +d*x**2*y -c*y**3/3;
                3        2       2     2  2      2
          2*C*Y - 6*D*X *Y -3*PX - 3*PY -3*X - 3*Y
   H  := - ------------------------------------------
                            6
g:= x**4 + 4*x**2*y**2 - 4*px*(px*y - py*x) + 4*x**2*y +
      3*(px**2 + x**2);
             2       2           4     2 2      2
   G  := -  (4*PX *Y - 3*PX - 4*PX*PY*X - X - 4*X *Y - 4*X *Y
         2
       -3*X )
depend x,t$
depend y,t$
depend px,t$
```

```
depend py,t$
gg:= df(g,t);
  GG := - 2*(4*DF(PX,T)*PX*Y - 3*DF(PX,T)*PX - 2*DF(PX,T)*PY*X
                                                           3
  - 2*DF(PY,T)*PX*X - 2*DF(X,T)*PX*PY - 2*DF(X,T)*X  -
               2                                                    2
  4*DF(X,T)*X*Y  -4*DF(X,T)*X*Y - 3*DF(X,T)*X + 2*DF(Y,T)*PX
               2                2
  - 4*DF(Y,T)*X *Y - 2*DF(Y,T)*X )
% Shows that the depend statements lead to gg in terms of
% df(x,t) and so on, before these derivatives are assigned
% explicit forms
df(x,t):= df(h,px)$
df(y,t):= df(h,py)$
df(px,t):= - df(h,x)$
df(py,t):= - df(h,y)$
% Hamilton's equations
gg:= gg;
                  2         2          2
  GG := 4*X*(C*PX*Y  - D*PX*X  + 4*D*PX*Y  - 3*D*PX*Y -
                  2         2
      2*D*PY*X*Y + PX*X  + 2*PX*Y  + 3*PX*Y + 2*PY*X*Y)
% This Poisson bracket form is not too helpful as it stands
factor(x,y,px,py);
gg:= gg;  % to clarify the structure of gg
               3                   2                          2
    GG := 4*X *PX*(- D + 1) + 8*X *Y*PY*(- D + 1) + 4*X*Y
        *PX*(C + 4*D + 2) + 12*X*Y*PX*( - D + 1)
d:= 1$
c:= - 6$
gg:= gg;
   GG := 0
end;
```

Since we may well generate an intermediate output of many lines in the course of a REDUCE calculation, so that its essential structure may be obscure, the ability to present output in different ways is very useful. While the command factor x,. . is used for a similar task to those carried out by the switch factor, it is not initiated by on, and cannot be cancelled by off. To cancel

```
factor x,y,px,py;
```

we must use

```
remfac x,y,px,py;
```

Factor commands of this type are additive. Suppose we wish to examine various forms of gg in this program, first grouping terms which are like with regard to x, y only, then with regard to x, y, and also px, py, and finally putting all the D terms together. We use this sequence

```
factor x, y;
gg:= gg;
factor px,py;
gg:= gg;
remfac x,y,px,py;gg:= gg;
factor D;
gg:= gg;
```

Another command used to rearrange terms is

```
order x,y;
```

which causes terms with x to appear first, ordered by power, then those without x but with y, then those without either. Applied to gg this gives

$$GG := 4*X*(-X^2 *D*PX + X^2 *PX - 2*X*Y*D*PY + 2*X*Y*PY +$$

terms without X)

The command order is cancelled, and a different one specified, by

```
order nil;
order C,D; gg:= gg;
```

which gives

$$GG:= 4*X*(C*PX*Y^2 +$$

all terms with factor D + all remaining terms)

We also use program 1.1 to illustrate that one aspect of the command

```
df(x,t):= df(h,px);
```

can be separated out by first using

```
depend x,t;
```

and similar commands for the other dependent variables. When these are given, but before the explicit right hand sides of the Hamilton's equations have been provided, the time derivative dG/dt is given in the form

$$\partial G/\partial x \cdot dx/dt + \partial G/\partial y \cdot dy/dt + \partial G/\partial p_x \cdot dp_x/dt \\ + \partial G/\partial p_y \cdot dp_y/dt. \tag{1.5}$$

Here the partial derivatives have been evaluated but the time derivatives remain in the form $df(x,t)$. Thus the command depend means that x is to be treated as an unspecified function of t; this requirement can be cancelled by using

```
nodepend x,t;
```

Looking back at the output of program 1.1 we see the form (1.5) of dG/dt, and also the form

$$\{\partial G/\partial x \cdot \partial H/\partial p_x + \partial G/\partial y \cdot \partial H/\partial p_y - \partial G/\partial p_x \cdot \partial H/\partial x \\ - \partial G/\partial p_y \cdot \partial H/\partial y\} \tag{1.6}$$

which is known as the Poisson bracket of G and H. The Poisson bracket is obtained by using Hamilton's equations (1.3), in the specific form (1.4), for the time derivatives of the variables x, y, p_x, p_y in (1.6). We also see that three terms in this Poisson bracket vanish for $D = 1$, leaving a term which vanishes, given this value of D, for $C = -6$.

1.6   CHANGES OF VARIABLE

Our next example, program 1.2, illustrates how changes of variable can be an important tool for exposing the structure of a problem. The intention here is to find another set of values of C, D for which the system is simply described, this time by decoupling the Hamilton equations. As already mentioned, for $D = C = 0$ the Hamiltonian (1.2) is that of two linear oscillators, which have the same frequencies since the masses are equal and the restoring forces are equal. This is a separable system for which the Hamilton equations can be decoupled. There are two constants of motion, the energies of the x-motion and the y-motion. We can easily identify normal modes, by changing variables to $u = x + y$, $v = x - y$. Each of these satisfies the same oscillator equation,

$$\mathrm{d}^2 u/\mathrm{d}t^2 = -u \qquad \mathrm{d}^2 v/\mathrm{d}t^2 = -v.$$

When we take nonzero values of D and C, there is no question of separating equations (1.4) in terms of x and y. However it is possible that we can still separate in terms of u and v, for particular values of D and C. The reason is that the kinetic energy terms are still simple, so that we can readily generate equations for the u and v accelerations. The question is, are the force terms in these equations respectively functions only of u and only of v, for specific C and D values?

Program 1.2 treats this problem in a straightforward way. We go from $H(x, y)$ to $H(u, v)$, using

```
let x = (u+v)/2;
let y = (u-v)/2;
```

We then compute new force terms, and integrate them to obtain potentials. This process is only correct when the potentials are respectively functions of u only and of v only. We check by eye that both are of the correct form if $C = -D$. We leave to an exercise in a later chapter the step needed for a fully automatic calculation, involving isolating the unwanted parts and solving an equation for C/D. A consequence of separability is again that the system has two constants of motion, the energy associated with the relative coordinate v and that associated with the centre of mass coordinate u. Our program verifies this.

PROGRAM 1.2

```
% Tests the use of let x = (u+v)/2 and let y = (u-v)/2 to
% separate a special case of the Henon--Heiles Hamiltonian
h:= (x^2 + y^2 + pu^2 + pv^2)/2 + D*x**2*y - C*y**3/3$
let x = (u+v)/2;
let y = (u-v)/2;
df(u,t):= df(h,pu);
df(v,t):= df(h,pv);
df(pu,t):= -df(h,u);
df(pv,t):= df(h,v);
pot1:= - int(df(pu,t),u);
```

$$POT1 := - (U*(C*U^2 - 3*C*U*V + 3*C*V^2 - 3*D*U^2 - 3*D*U*V + 3*D*V^2 - 6*U))/24$$

```
pot2:= - int(df(pv,t),v);
```

$$POT2 := (V*(3*C*U^2 - 3*C*U*V + C*V^2 + 3*D*U^2 - 3*D*U*V - 3*D*V^2 + 6*V))/24$$

```
% potentials depend on both variables
D:= 1$
C:= - 1$
pot1:= pot1;
```

$$POT1 := \frac{U^2*(2*U + 3)}{12}$$

```
% this potential now depends only on u
pot2:= pot2;
```

$$POT2 := - \frac{V^2*(2*V - 3)}{12}$$

```
% this potential now depends only on v
try1:= df(pu**2/2 + pot1, t);
    TRY1 := 0
```

```
try2:= df(pv**2/2 + pot2, t);
    TRY2 := 0
% the energy has two constant parts
end;
```

1.7   SETTING VALUES LOCALLY AND GLOBALLY

In program 1.2 we change variables by the use of the command let, for example

```
let x = (u+v)/2;
```

An alternative is to use

```
x:= (u+v)/2;
```

and so on. In this program the consequences are exactly the same, but this is not true in general. A safe general principle is to use := to assign a value to a newly introduced variable, and let . . = . . to assign a value to an expression which is given in terms of variables already present. We give some examples to illustrate the different effects of these commands. Suppose we have two variables x and y, and we wish to identify x**2 with y. If y is clear and stays clear (its value is always y) and also x is clear throughout,

```
x**2:= y;
```

and

```
let x**2 = y;
```

have the same effect.   But if y is assigned different values, the consequences for x**2 are different, as in this sequence, which starts by assigning a value to y,

```
y:= v;
    Y := V
x**2:= y;
     2
    X   := V
y:= w;
    Y := W
x**2;
    V
```

and this, starting in the same manner,

```
y:= v;
   Y := V
let x**2 = y$
x**2;
   V
y:= w;
   Y := W
x**2;
   W
```

or this one

```
x:= a+b;
   X := A+B
let x**2 = y$
square:= x**2;
               2           2
 SQUARE := A + 2*A*B + B
clear x$
y:= v;
   Y := V
sq:= x**2;
   SQ := V
square;
   2           2
 A + 2*A*B + B
```

So the command x**2:= y assigns to x**2 the current value of y, and this value is retained by x**2 when y is subsequently assigned another value. The command let x**2 = y means that x**2 follows any changes in the value of y, so long as x is clear (has value x). If x is later assigned a value then x**2 must conform with this value. The effect of the command let is cancelled by

```
clear x**2$
```

Suppose that y is clear when we set x:= y, and then takes two successive values. The sequence is now

```
x:=y;
   X := Y
y:=1;
   Y := 1
x;
   1
y:=2;
   Y :=2
x;
   2
```

Suppose that one inadvertently uses x**2 = y; in place of either x**2:= y; or let x**2 = y. This does not elicit an error message, and will give the output

```
   2
X    = Y
```

but the value apparently assigned is not retained. The = sign on its own is to be avoided. Its proper place is after let, or in other particular contexts which we shall explain later, such as in the expression sub(x = 0, y) (chapter 2) or in the output of the command solve (chapter 3).

Mathematically there is no distinction between the pairs of equations

$$x = (u + v)/2 \qquad y = (u - v)/2$$

and

$$x + y = u \qquad x - y = v$$

but as assignments these are very different, whether := or let is used. The first pair are unambiguous. In the second pair, first x is interpreted as u - y, and then as v + y, and the second interpretation overrides the first. It is vital to avoid ambiguity whenever changes of variable are employed.

Summary of commands and conventions introduced in this chapter

```
cue number      1:
terminators    ;  $
a:= b      let a = b      clear a        a; or a:= a;
on/off factor
```

```
df(a,x)    df(a,x,n)     int(a,x)
bye        pause     cont       end   on/off demo
ws(n)
in/out filename
depend x,t     nodepend x,t
factor x,y, . .    remfac x, y, . .
order  x, y, . .   order nil
comment    text  $            %      text
```

2

Some basic tasks

In which we discuss matters of arithmetic, introduce several new switches, and investigate what REDUCE knows about elementary functions, and the operations of differentiation and integration.

2.1 INTEGER AND DECIMAL ARITHMETIC

In this chapter we introduce a number of rather basic aspects of REDUCE. Instead of a case study we find it more appropriate to present a large number of short hands-on sequences, again adopting the convention of ignoring cue numbers. First we look at the variety of types of arithmetic that REDUCE can perform. REDUCE normally performs exact integer arithmetic. If the input includes a number in decimal form, it is interpreted as an approximate ratio of integers,

```
a:= 0.5;
  A:= 0.5
c:= a*b;
***  0.5 represented by 1/2
      B
  C:= ---
      2
```

However, even the relatively simple ratio b:= 0.55 can be represented, not by 11/20, but by the ratio of two 16-figure numbers, if this translation is left to the whim of the system. So one should avoid entering numbers in this form unless they are to be treated as decimals.

The natural mode of operation of REDUCE is with ratios of integers, and unless one has good reason, information given in decimal terms should be translated before entering it. One can get 6-figure accuracy from a ratio of two 3-figure integers.

Input and output can be interpreted in decimal form by using a switch, on float.

```
a:= 1/2;
      1
 A:= ---
      2
on float;
a:=a;
 A:= 0.5
```

For example, a result expressed as the ratio of two large integers is more readily understood if translated in this way. Again, if we ask REDUCE to calculate a transcendental function, except for a few special arguments we get this sort of result

```
s:= sin(1/2);
            1
   S:= SIN( - )
            2
```

whereas with two switches on

```
on float, numval;
s:= sin(1/2);
   S:= 0.4794255386042029
```

There is an obvious defect here; the result has an arbitrary, and excessively large, number of significant figures. To cure this we use the switch bigfloat instead of the switch float; it is called this because we could, if we wished, set an even more excessive precision. However after bigfloat (but not after float) we can set n significant figures, as in this sequence

```
on bigfloat;
precision 5;
5
on numval;
s:= sin(1/2);
   S:= 0.47943
```

Notice the slightly different form of output here; after entering precision n, the number of significant figures is given at the left.

EXERCISE 2.1 Calculate $\sqrt{2}$ (hands on) using the elementary iterative process $x_1 = 3/2$, $x_2 = (2/x_1 + x_1)/2$, $x_3 = (2/x_2 + x_2)/2, \ldots$, starting with integer ratio arithmetic and converting to floating point after the fourth step.

2.2 COMPLEX AND MODULAR ARITHMETIC

REDUCE can also do complex arithmetic. The notation I is reserved for the square root of -1, with I^2 automatically replaced by -1. So we must avoid using I as the name of a variable. The choice of operations available in complex arithmetic can be extended by using some more switches. Thus, while the sequence

```
a:= x + i*y$
b:= x - i*y$
c:= a*b;
        2    2
    C:= X  + Y
```

requires only replacing I^2 by - 1, by using a switch, on complex, one can search a real expression for complex factors, as in

```
on complex;
a:= x**2 + y**2$
b:= x - i*y$
c:= a/b;
    C:= X + I*Y
```

Another switch, on rationalize, allows us to reduce the ratio of two complex numbers to a single complex number, as in

```
on rationalize;
a:= x + i*y$
b:= x - i*y$
c:= a/b;
```

$$C := \frac{2*I*X*Y + X^2 - Y^2}{X^2 + Y^2}$$

We can also reverse this process, using the switch on gcd, as in

```
on complex, gcd;
a:= x**2 - y**2 + 2*i*x*y$
b:= x**2 + y**2$
c:= a/b;
```

$$C := \frac{X + I*Y}{X - I*Y}$$

This switch causes a search for common factors in numerator and denominator, whether complex numbers are involved or not. Another situation in which the symbol I will appear is in using the square root function, sqrt(a). Thus we have

```
sq:= sqrt(n**2);
    SQ:= N
```

and

```
sq:= sqrt(-n**2);
    SQ:= I*N
```

There is a potential source of trouble here if we should later assign n a negative value. We could however use the sequence

```
on precise;
sq:= sqrt(-n**2);
    SQ:= I*ABS(N)
```

where the function abs(x) is the modulus, or absolute value, of x. Another switch, on reduced, allows the sequence

```
on reduced;
sq:= sqrt(a*b);
  SQ:= SQRT(A)*SQRT(B)
```

whereas without this switch we have

```
sq:= sqrt(a*b);
   SQ:= SQRT(A*B)
```

Here the switch is of a rather different type to those used previously. With factor on, any expression is searched for factors, but with reduced on only expressions operated on by sqrt are put in a specific form. Once evaluated, these retain their form even when reduced is off, as in

```
on reduced;
sq:= sqrt(a*b);
   SQ:= SQRT(B)*SQRT(A)
off reduced;
sq:= sq;
   SQ:= SQRT(B)*SQRT(A)
qs:= sqrt(a*b);
   QS:= SQRT(A*B)
```

Modular integer arithmetic can be performed by first setting the desired modulus, for example by

```
setmod 5;
```

and then setting a new switch,

```
on modular;
```

The switch does not affect any integer exponent. Thus we have

```
setmod 5;
on modular;
x:= 6*y - 2*y**6;
              6
   X:= Y + 3*Y
```

Note that -2 is -5+3, not -(0+2).

While all the switches introduced so far are normally off (the default state is off) some are normally on. For example, suppose we return to our very first sequence in chapter 1, but start by setting

```
a:= 2+ 4*x + 2*x**2;
          2
   A:= 2*(X + 2*X + 1)
```

We did not ask for the numerical factor to be isolated. To remove it we use

```
off allfac;
a:= a;
        2
  A:= 2*X + 4*X + 2
```

The maximum and minimum functions can be used for numerical arguments only. Here for example max(1,2,3), and min(1,-1) evaluate to 3, -1, respectively.

2.3 ELEMENTARY TRANSCENDENTAL FUNCTIONS

When we use a transcendental function such as $\sin(x)$ in REDUCE, we must be clear what we can take for granted that REDUCE 'knows' about this function, and what we have to specify explicitly. We have seen that in general $\sin(x)$, with a numerical value of x, will give a decimal value when the appropriate swiches are on. For certain special values of x, namely zero, integer multiples of $\pi/4$ and of $\pi/6$, the sine is given in terms of integers and sqrt(2), sqrt(3). The notation PI is reserved for π, and must not be used as a variable name. Also $\sin(-x)$ is given as $-\sin(x)$. Similar special cases are treated in this way for $\cos(x)$. With other functions the choice is more limited, for example $\tan(0) = 0$, $\exp(0) = 1$, $\log(1) = 0$. The notation E is reserved for e, the base of natural logarithms, and we have

```
x:= exp(y);
        Y
   X:= E
z:= e**5;
        5
   Z:= E
l:= log(e);
   L:= 1
lz:= log(z);
   LZ:= 5
lx:= log(x);
   LX:= Y
```

However these few results, together with the ability to differentiate and integrate, are all we can rely on. If we wish to use such basic results as

$$\sin^2(x) + \cos^2(x) = 1$$
$$\log(xy) = y\log(x)$$
$$\exp(\mathrm{i}x) = \cos(x) + \mathrm{i}\sin(x)$$

we must provide them, as `let` statements. Just as we found that `let x + y = u` is interpreted as 'assign to x the value $u - y$', so we have to find by experience how REDUCE interprets the command

```
let sin(x)**2 + cos(x)**2 = 1;
```

and whether this differs from the interpretation of

```
let cos(x)**2 + sin(x)**2 = 1;
```

or of

```
let sin(x)**2 = 1 - cos(y)**2;
```

If this sum of squares appears explicitly in an expression, there is no problem, but given a polynomial in $\cos^2(x)$ and $\sin^2(x)$, will it be turned into a polynomial in $\cos(x)$ alone, or in $\sin(x)$ alone? And does $\sin(x)$ finish up as a square root? Needless to say, further useful results for $\cos(2x)$, $\sin(3x)$ and so on have to be introduced with due care. This topic is resumed in chapter 7.

EXERCISE 2.2 Set up some expressions in $\sin(x)$ and $\cos(x)$ and investigate the use of the sum of squares rule.

Now suppose that we have an expression containing squares of $\cos(x)$, $\sin(x)$ and also of $\sin(y)$ and $\cos(y)$. We do not have to give the sum of squares rule for both angles, we can instead use a more powerful kind of instruction,

```
for all x let sin(x)**2 + cos(x)**2 = 1;
```

We can also use, for example,

```
for all x, y let cos(x+y) = cos(x)*cos(y) - sin(x)*sin(y);
```

As for the simple let, it is vital to avoid ambiguity when a number of these commands are used.

EXERCISE 2.3 Investigate whether the command just mentioned will deal with the case of $x = y$, that is, of $\cos(2x)$; if not, devise a command that does. Also investigate whether the command

```
for all x let exp(i*x) = cos(x) + i*sin(x);
```

suffices to convert $\exp(ix) + \exp(-ix)$ into $2\cos(x)$.

2.4 INTEGRATION

In REDUCE we can apply the familiar rules for differentiating a sum, product or ratio of two functions, and so we can differentiate functions built up from elementary transcendental functions. We have already seen in chapter 1 the use of the chain rule, for differentiating a function of a function. These four rules, together with the ability to apply them to more than one independent variable, suffice to differentiate. It is a crucial distinction between differentiation and integration, as taught in conventional calculus texts and courses, that the first proceeds in an algorithmic manner but the second relies on a collection of special tricks. Even the additive rule for integration has to be regarded with caution, since it is quite possible for $f(x) + g(x)$ to have an indefinite integral which can be evaluated, although $f(x)$ and $g(x)$ separately do not. Computer algebra requires a radical reappraisal of integration, as discussed for example by Davenport *et al* (1988), because it seeks to apply algorithms.

We do not go into this, but merely illustrate some aspects of the procedure and some successes and failures. REDUCE attempts indefinite integration with respect to x, of an explicit function $y(x)$, when we give the command z:= int(y,x); or ask for the expression int(y,x). Constants of integration are not put in explicitly. Definite integrals can be performed if and only if the indefinite integral is obtained explicitly, by then using the substitution expression sub(x=a, z). Thus to integrate

$4x^3$ from $x = 1$ to $x = 3$ we use the sequence

```
y:= int(4*x**3,x);
        4
    Y:=  X
y13:= sub(x=3, y) - sub(x=1, y);
    Y13:= 80
```

The substitution expression is quite different from a command that assigns a value to x. The x in sub(x=a, y) applies only to the x appearing in the expression y, and only at the point of use; otherwise we would be unable to use sub(x=a,y) twice, with a = 1 and a = 3, in the example above. Another, less generally useful, kind of substitution is carried out by the where operator, as in

```
x**2 where x = y;
   2
   Y
x**2 + y**4 where x = y, y = x;
   2     4
   Y  +  X
```

This second example does not, as one might have suspected, lead to an infinite regression.

Returning to integration, we cannot use the value infinity for x in the substitution expression. With due care to satisfy ourselves that the indefinite integral is zero in the limit of x tending to infinity, we can still produce simple integrals from zero to infinity by using

```
y:= - sub(x=0,y)$.
```

The real limitation of REDUCE in definite integration is that there are many known integrals in the range $(0, \infty)$ or $(-\infty, \infty)$ for which the corresponding indefinite integral is not known. Such integrals are often obtained by path integration in the complex plane. REDUCE does not allow access to such definite integrals.

We turn now to the question of how wide a variety of indefinite integrals REDUCE can perform. We illustrate this by two lists, of successes and failures. Among the successes are some, involving square roots, for which success depends on using a user-contributed package,

ALGINT, due to J H Davenport, and supplied with REDUCE 3.3 or 3.4. These are labelled ALGINT in the list. ALGINT does not deal with integrals combining a square root and a transcendental function. Failures are signalled either by returning the input as output, thus

```
i1:= int(cos(x)*atan(x),x);
   I1:= INT(COS(X)*ATAN(X),X)
```

or by an apparently unending search, which we can break off by an appropriate interrupt command.

SUCCESSES

$\mathrm{erf}(x)$

$\mathrm{atan}(x)$ \qquad $\mathrm{atanh}(x)$

$\mathrm{dilog}(x)$

$\exp(x)\cos(x)$ \qquad $\exp(x)\cos(x)(x^2+1)$

$1/(x^2+1)$ \qquad $1/(x^2-1)$

$1/((ax+b)(cx+d))$

$x^3\,\mathrm{dilog}(x)$

$\mathrm{sqrt}(ax+b)$ \qquad uses ALGINT

$\mathrm{sqrt}(x+\mathrm{sqrt}(x^2+1))/x$ \qquad uses ALGINT

FAILURES

$\exp(x)\log(x)$

$\cos(x)*\mathrm{atan}(x)$

$\log(x)\sin(x)$

$\sin(x)/(x^2+1)$

$\cos(x)/(x^2-1)$

$\exp(x)/x$

$\exp(x)/\sin(x)$

$\sin(x)/(ax+b)^2$

$\mathrm{sqrt}(x^2+1)/\mathrm{sqrt}(x^2-1)$

EXERCISE 2.4 For equations of the Lienard type (generalised van der Pol oscillator equations)

$$d^2x/dt^2 + dx/dt f(x) + kx = 0$$

the stability of a periodic solution can be determined from the sign of the integral

$$I = \int_0^T f(x(t)) dt$$

where T is the period. (See for example Mickens (1981).) Now in chapter 7 we shall study a method for obtaining approximate periodic solutions. The first approximation is simply

$$x_0(t) = A \cos(pt)$$

with $p^2 = k$, and A depending on parameters used in $f(x)$. For certain equations of this type, used as models of oscillatory circuits, $f(x)$ takes the form

$$(a + bx^2)/(c + dx^2).$$

Examples of such equations are due to Scott (1968) and to Walker and Connelly (1983); the second of these is given in chapter 7, equation (7.8).

Show that for this form of $f(x)$ the approximate integral

$$I_0 = \int_0^{2\pi/p} f(x_0) dt$$

can be determined by REDUCE without first providing the information that, for all x,

$$\cos(x)^2 = (1 + \cos(2x))/2$$

but that the process is faster if one does provide this information. (The switch on time can be used to yield an extra output line giving time taken for each output.)   What other information about trigonometric functions has to be provided once the definite integral is evaluated? Try also the form, found in another such model, the Ceschia–Zecchin (1981) equation,

$$f(x) = (a + bx^2 + cx^4)/(d + fx^2 + gx^4).$$

The use of an approximate stability integral is of course not to show that the limit cycle is 'approximately stable'; it is to indicate approximately the sensitivity of stability to the parameters that occur in $f(x)$.

Summary of commands and conventions introduced in this chapter

```
on/off  float   numval  bigfloat   complex   rationalize
        gcd    reduced   precise   modular
off/on allfac

precision n (after on bigfloat)
setmod m (before on modular)

abs(x)
sub(x=a,y)    for all x let
where x = y
on/off time
```

I is the square root of -1. PI is π. E is the base of natural logarithms, e. Transcendental functions (of a single real variable) available are sin cos   tan   cot   asin   acos   atan   exp   log   sinh cosh tanh   asinh   acosh   sqrt   dilog   erf   expint

Additions and amendments for users of 3.4

The new 3.4 edition of REDUCE handles floating point arithmetic in a different manner. The switch on rounded is used, combined with the command precision 10;, for example. However precision 0; gives the precision normal for the computer environment used. The command print_precision 5;, for example, makes the output more compact. A program using float and bigfloat will run in 3.4, but it would be safer to convert to the new method.

3.4 gives a variety of other operations on variables taking numerical values,

`ceiling(x)` next integer above x

`floor(x)` integer immediately below x

`conj(x)` complex conjugate of x

`fix(x)` integer part of x (= `floor` for $x > 0$, = `ceiling` for $x < 0$)

`round(x)` nearest integer to x

`impart(x)` imaginary part of x

`repart(x)` real part of x

For variables taking integer values,

`factorial(x)`

`nextprime(x)`

An alternative, and more flexible, formulation for substitutions has been introduced in 3.4. Instead of

```
for all x let cos(x)**2 + sin(x)**2 = 1;
```

one can use a rule

```
cos(~x)**2 + sin(~x)**2 => 1;
```

Two new notations are introduced, the tilde $\sim$ before the free variable on the left, and the => replacing =. Rules can be set out in lists, which can be given names

```
trig1:= {cos(~x)**2 => (1+cos(2*x))/2,
         sin(~x)**2 => (1-cos(2*x))/2};
```

Then one can apply the rules globally with

```
let trig1;
```

and remove them with

```
clearrules trig1;
```

Alternatively rules can be applied locally using where, for example as

```
prodab:= triga*trigb where {cos(~x)**2
=> (1+cos(2*x))/2,
sin(~x)**2 => (1-cos(2*x))/2};
```

(Here `triga` and `trigb` would be previously defined to involve powers of $\sin(x)$ and $\cos(x)$.) It is precisely this power to apply rules either locally or globally that gives greater flexibility, combining features of `for all . . let` with features of `sub`. Rules can also be qualified as in

```
operator posfilter;
posfilter(~n) => 0 when n < 0;
posfilter(~n) => 1 when n > 0;
```

(Notice that this is 'when' not 'where'.)

3

Linear   algebraic   equations

In which we examine the command `solve`, learn about lists, and illustrate by constructing a program for dimensional analysis.

3.1   THE SOLVE COMMAND AND LISTS

This chapter is primarily concerned with the command `solve`, which allows us to solve a linear, quadratic, cubic or quartic algebraic equation, or a set of simultaneous linear equations. The trivial process of solving one linear equation is exactly what we need to complete program 1.2 of chapter 1, by automatically identifying the ratio C/D. Even in this case we need to look at how the result is presented, and how it is put into the next step of the calculation. So we look first at the sequence

```
f:= a + x;
  F:= A + X
sol1:= solve(f,x);
  SOL1:= { X = - A}
```

and at the sequence

```
quad:= a*x**2 + b*x + c;
            2
  QUAD:= A*X + B*X + C
sol2:= solve(quad,x);
   SOL2:=                  2
                - B + SQRT(B - 4*A*C)
       { X =  -----------------------   ,
                       2*A
                        2
                - B - SQRT(B  - 4*A*C)
        X =   -----------------------   }
                       2*A
```

The results are in each case presented as a list, signalled by the brackets { }. In sol1 this list is made up of one item, the solution for x. In sol2 the list is made up of the two alternative solutions for x.

This is all we need when the final result required is the solution, or the set of solutions, in the list. The results can be read off, and we need not be concerned with the fact that they make up a list. But we often have to use the values in a later step in the calculation, and so we need to know how to extract an expression from a list. To extract x from sol1 we use

```
x:= rhs first sol1;
   X:=  - A
```

Thus x is defined as the right hand side of the first member of the list sol1 (which, as it happens, has only one member). To make this quite clear, break it into two steps:

```
first sol1;
   X = - A
rhs first sol1;
       - A
```

The first three members of a list l1 can be identified as first l1, second l1, third l1; The nth member of a list l2, having at least n members, can be identified by part(l2,n). We extract the two values of x in the quadratic case by

```
x1:= rhs first sol2; x2:= rhs second sol2; .
```

EXERCISE 3.1 Complete program 1.2 by equating to zero the unwanted part of each of the potentials pot1 and pot2, and verifying that they give the same C/D.

EXERCISE 3.2 Solve the cubic equation

$$x^3 + px + q = 0$$

and examine ways of simplifying the roots by denoting certain repeated expressions by single symbols. Identify the root that is always real, and obtain a criterion for the other two roots to be real.

EXERCISE 3.3 The expression $y = (ax^2 + bx + c)/(fx^2 + gx + h)$. Use solve and the commands

```
ny:= num(y);
dy:= den(y);
```

which select the numerator and denominator, to find the values of x at which y is zero or infinity.

Of rather more interest is the use of solve to handle a set of simultaneous linear equations. Here the command solve is presented thus

```
list:= solve({eq1, eq2, eq3, . . }, x1, x2, x3, ..)
```

with the equations listed inside { }, but the unknowns without the { }. The output is in a form having {{ }}, a list within a list.

```
LIST:= {{ X1 = , X2 = , . . , XN = }}
```

(in a line for short expressions, otherwise in a column). Here first list now stands for everything within the outer { }, that is the list of solutions

```
{ X1 = , X2 = , , XN = }
```

and consequently the value of x1 is extracted by the command

```
x1:= rhs first first list;
```

that of x2 by the command

```
x2:= rhs second first list
```

and so on. However if we wish to substitute for x1 in an expression y, we need not identify x1 with an rhs; we use

```
sub(first first list, y);
```

Here the first equation in the inner output list is needed, rather than its right hand side, since the command sub is of the form

```
sub(p=q,r);
```

This illustrates the significance of the = operator mentioned briefly in chapter 1. In fact, if one or more of the xj appear in y, we can use

```
sub(first list, y);
```

to replace each of them by the corresponding solution. The irrelevant parts of the list are ignored.

There are several other operations that we can perform on lists, as in this sequence

```
l1:= {a,b,c,d}$  l2:= {f,g,h}$
length l1;
     4
append (l1,l2);
     {A,B,C,D,F,G,H}
reverse l2;
     {H,G,F}
rest l1;
     {B,C,D}
a.l2;
     {A,F,G,H}
l1.l2;
     {{A,B,C,D},F,G,H}
```

The . operation, also known as cons, is basic to the way LISP handles lists, in terms of the first part of a list, and the rest of the list. The append operation in

append (l1, l2);

is equivalent to a recursive use of cons, working from back to front of l1, adding terms at the beginning of l2. But l1.l2 adds the list l1 as the first element in a new list, of which the rest contains the three elements of l2.

3.2 DIMENSIONAL ANALYSIS

Dimensional analysis, in its simplest manifestation, is a technique for establishing a power law for a certain physical quantity, which we shall call the target, in terms of other relevant physical quantities. For a modern book on the subject see Barenblatt (1987); Pankhurst (1964) is an old book with numerous explicit examples.

Dimensional analysis provides an interesting example in which the essence of the calculation is the solution of a set of simultaneous linear equations. It is interesting because we can exploit the flexibility of the command solve, shown in the fact that it can cope with nonstandard sets of unknowns. In the standard case there are n inhomogeneous equations with n unknowns, and all the equations are linearly independent. Using matrix notation, such a set of equations is

$$\mathbf{A}X = U \tag{3.1}$$

where $\mathbf{A}$ is an $n \times n$ matrix of coefficients, having rank n, X is the $n \times 1$ matrix (vector) of the x_j, and U is a vector of n parameters u_j, independent of the unknowns. The solution is a set of n explicit expressions for the unknowns. In our program for dimensional analysis it is convenient to solve a fixed number of equations for a fixed number of unknowns. But in a specific application one or more equations may be left out, or there may be one or more superfluous unknowns, which can take arbitrary values. More significant is the fact that there may be a subset of the unknowns which can be determined only to within one arbitrary parameter. The solve command can handle this case without the need for prior identification of the special nature of the subset.

Dimensional analysis, we repeat, is the search for a power law for a target physical quantity in terms of a number of other relevant physical quantities. Choosing which physical quantities are relevant requires some understanding of the physical problem. Assuming a power law relation

$$P_T = P_1^{a_1} P_2^{a_2} \ldots P_n^{a_n} \tag{3.2}$$

the physical dimensions on either side of this relation must match. The modern convention is to use five basic dimensions, with corresponding base units in the SI convention; three are mechanical

mass [M]	kilogram
length [L]	metre
time [T]	second

and the others are

current [I]	ampere
temperature [O]	degree Kelvin

The other two SI base units, mole and candela, are less likely to occur. Other choices of basic dimension are found in older texts, in particular for the electromagnetic dimension, and sometimes force is used in place of mass.

Now the target P_T and the other quantities P_j each have specific dimension. For example, if the target is the rotational energy of a rigid body, and the relevant quantities are velocity, length and moment of inertia, we have $n = 3$ in equation (3.2), and we can set down dimensional equations

$$\begin{aligned} P_T &= [M][L]^2[T]^{-2} & P_1 &= [L][T]^{-1} \\ P_2 &= [L] & P_3 &= [M][L]^2. \end{aligned} \tag{3.3}$$

We use the equality sign in a rather special way here, meaning for example that any velocity is a distance divided by a time. The assumed power law is only meaningful if, for each dimension, the sum of the powers on the right hand side of equation (3.2) is equal to the power of this dimension in the target quantity. Thus in the example (3.3) this must apply for each of the three mechanical dimensions, giving

$$1 = a_3 \qquad \text{for mass}$$
$$2 = a_1 + a_2 + 2a_3 \qquad \text{for length}$$
$$-2 = -a_1 \qquad \text{for time}$$

Thus the equation (3.1) is

$$\begin{pmatrix} 0 & 0 & 1 \\ 1 & 1 & 2 \\ -1 & 0 & 0 \end{pmatrix} \begin{pmatrix} a_1 \\ a_2 \\ a_3 \end{pmatrix} = \begin{pmatrix} 1 \\ 2 \\ -2 \end{pmatrix}$$

with the solution $a_1 = 1$, $a_2 = -2$, $a_3 = 1$. We interpret this result to mean that the required energy is

$$P_T = c \cdot P_1^2 \cdot P_2^{-2} \cdot P_3$$

where c is an arbitrary numerical factor. Here the rotational energy is proportional to the moment of inertia and to the square of the angular velocity.

So the problem is one of solving simultaneous linear equations to find the unknown powers. These equations are inhomogeneous, that is some or all of them have a term not containing one of the unknowns (an element of the vector U in equation (3.1)). There are at most five equations, and our program assumes that there are five. Any superfluous equations must be listed after the others; then they are ignored by solve. The main complication is the presence of dimensionless groups. A dimensionless group is a set of physical quanties for which a product of powers can be found which is dimensionless. A trivial example is a ratio of two characteristic times. A slightly subtler example is a force F, $[M][L][T]^{-2}$, a linear density D, $[M][L]^{-1}$ and a velocity V, $[L][T]^{-1}$, with the product $FD^{-1}V^{-2}$ having dimension zero.

If the reader has not encountered dimensional analysis, he may be dubious of its value. After all, there are many ways in which physical quantities are related other than by power laws. The concept of a dimensionless group offers an answer to this objection. If the target depends on some physical quantities by way of an exponential, for example, then these have to occur in a combination that is dimensionless, since the argument of an exponential is a pure number. In dimensional analysis these dimensionless combinations are found to appear to an arbitrary power. This implies that they can occur as an arbitrary power series (analytic function). In fact identifying relevant dimensionless groups is one of the most useful aspects of dimensional analysis.

Pankhurst (1964) gives a simple example, with an obvious dimensionless group. This concerns the volume rate of flow Q of a liquid of viscosity v along a narrow tube of radius a and length l, along which there is a pressure difference P. Dimensional analysis tells us (see exercise 3.4) that Q can be expressed in the form

$$Q = a^3 P v^{-1} (a/l)^n. \tag{3.4}$$

Here n is an arbitrary integer. This implies that Q can be expressed with an arbitrary power series present as a factor,

$$Q = a^3 P v^{-1} [c_0 + c_1(a/l) + \ldots$$

Consequently Q can be expressed in the form

$$Q = a_3 P v^{-1} F(a/l)$$

where F stands for an arbitrary function. Expressed another way, dimensional analysis tells us that apart from the obvious dimensionless group a/l there is another, Qv/Pa^3, and that there is some functional relationship between them,

$$Qv/Pa^3 = F(a/l).$$

If we wish to compare the flow of the same liquid in two tubes of different length, but with the same ratio a/l, at a constant pressure difference, this tells us that the flows are in the ratio of the radii cubed. There are many such scaling applications, usually modelling something large by a scaled down model, as in the use of wind tunnels in developing an aircraft design.

We do not have to notice the presence of a dimensionless group before applying solve; the group signals its presence in the output.

The program is constructed in one piece, so that different examples can be studied by editing the program input file. The program could readily be separated into a part that loads a specific problem, and a part that sets out and solves the simultaneous equations. In view of our comments in chapter 2 about the limited knowledge REDUCE has about elementary functions, it should not be surprising that the program starts by setting up the exponents in equation (3.2) as linear functions of the dimensions.

In program 3.1 we take an example in which there is one dimensionless group, one superfluous equation, and five relevant physical quantities. In the output the presence of the dimensionless group is signalled by the appearance of an arbitrary complex number ARBCOMPLEX(1) in AA and BB, AA also having a non-arbitrary term. This reveals that the ratio of PA and PB is the dimensionless group.

PROGRAM 3.1

```
% Dimensional analysis in terms of dimensions ma, len,
% tim, cur, and tem.
% Target = product of powers of up to 6 relevant
% physical quantities.
% The dimensional equations are logarithmic, so that
% we mean, for example, charge = current.time when we
% write pc:= cur + tim below.
% First part allows use of force as a mechanical dimension
% instead of mass, epsilon0 or mu0 as electromagnetic
% dimension instead of current. The list could be extended
% as a dictionary of dimensioned quantities.
force:= ma + len - 2*tim$
eps:= 2*cur + 4*tim - ma - 3*len$
mu:= ma + len - 2*cur - 2*tim$
% Target is energy radiated by oscillating charge.
% pa is wavelength, pb amplitude, pc charge, pd velocity
% of light, pe is vacuum permittivity epsilon0.
pa:= len$ pb:= len$ pc:= cur + tim$ pd:= len - tim$
pe:= eps$ target:= 2*len + ma - 3*tim$
% Arbitrary powers assigned, aa means power to which pa
% is raised.
s:= aa*pa + bb*pb + cc*pc + dd*pd + ee*pe - target$
```

```
% We use a quick way to find the power to which mass, for
% example, is raised when s is expanded.
cma:= df(s,ma);
   CMA:= - (EE + 1)
clen:= df(s,len);
   CLEN:= AA + BB + DD - 3*EE - 2
ctim:= df(s,tim);
   CTIM:= CC - DD + 4*EE + 3
ccur:= df(s,cur);
   CCUR:= CC + 2*EE
ctem:= df(s,tem);
   CTEM:= 0
solist:= solve({cma,clen,ctim,ccur,ctem},aa,bb,cc,dd,ee,ff);
   SOLIST := {{AA = (ARBCOMPLEX(1) + 2),
     BB = ARBCOMPLEX(1),
     CC = 2,
     DD = 1,
     EE = -1,
     FF = ARBCOMPLEX(2) }}
end;
```

This example is concerned with the rate of energy radiation from an electromagnetic oscillator. The first two physically relevant variables are lengths, the wavelength and the amplitude, so that we already know a dimensionless group. Then there are the charge, the velocity of light, and the vacuum permittivity ϵ_0. Our program allows up to six such quantities, and the sixth appears in the solution as the only term with ARBCOMPLEX(2). The five equations to be solved are given in terms of the powers of four dimensions actually present, and a dummy equation for temperature, which is not needed in this example,

```
solist:= solve({cma,clen,ctim,ccur,ctem}, aa,bb,cc,dd,ee,ff);
```

The output is a list, made up of a single list, which in turn is made up of six parts, each of which specifies one of the unknowns. Five of these are significant, but there are only four real powers, because ARBCOMPLEX(1) appears in two of the significant indices given by solve (those for the lengths). The dummy index is ARBCOMPLEX(2).

EXERCISE 3.4 Carry out the dimensional analysis of the example cited from Pankhurst (1964) above, leading to equation (3.4), and of the following examples:

The critical wind velocity for 'white caps' to form on the surface of deep water depends on the viscosities v_w and v_a and densities d_w and d_a of water and air respectively, and on the gravitational acceleration g. (In both these examples we use the dynamic viscosity, which has the dimensions $[M][L]^{-1}[T]^{-1}$.)

The vibration period of a magnet in a magnetic field, the relevant physical quantities being moment of inertia, magnetic induction and magnetic moment. Since our program contains dimensions of the vacuum permeability μ_0, the best way to remember the dimensions of magnetic quantities is from

> magnetic moment = coil area times current

and from

> line integral of induction round closed loop
> = current times μ_0.

EXERCISE 3.5 Complete these examples by putting the targets in the form of products of powers, and with their own names, to remove traces of the 'logarithmic' notation used. Are there any difficulties with ARBCOMPLEX(1) etc.?

Summary of commands and conventions introduced in this chapter

```
solve(a,x)   solve({a,b,  },x, y,  )
```

For any list $\{p, q, \ldots\}$

```
first list, second list, third list, part(list,n)
```

For list of form {x = a1, x = a2, . . }

```
rhs first list gives a1, etc.

sub(list,y)   sub(first list, y)
```

For list of form {{x=a, y=b, . .}}

`rhs first first list` gives a, etc.

When `solve` is given in terms of a list {{a,b, . }} parts of its output (called for example `solist`) are called up by

`rhs first first solist` and so on.

`reverse, . , rest, length` applied to list

`append` applied to two lists

```
num(y)    den(y)
arbcomplex(J)
cons (.)
```

4

Repetitive processes

In which we examine loops and procedures, illustrating with examples on Lagrangian equations and Poisson brackets, and learn something about matrices and about the truncation of series.

4.1 LOOPS AND PROCEDURES

In long calculations we are bound to encounter repetitive processes, and to seek concise ways to implement them without the need to type in repetitive instructions. Two situations are very common; a process is to be repeated several times in succession, or a process is to be available on call. The power to deal with these situations is of course a necessity in computer languages for numerical calculations also, and most readers will be familiar with how this is implemented in at least one language. Here we have to specify the particular ways these tasks are performed in REDUCE.

In the first situation we introduce an index or several indices, and compute some expression for a certain number of values of each index. This is a loop. The most commonly used loop command in REDUCE is

```
for K:= 1:n do     ;
```

which is short for

```
for K:=1 step 1 until n do     ;
```

After do there are one or more commands. If there are more than one, they are enclosed in carets << >>, and the last of these instructions has

no terminator. However there is a terminator after >>. For an example
consider the sequence

```
a:= 1$ b:= 1$
for K:=1 step 2 until 7 do <<
    a:= a*K; b:= b/K >>;
a;
    105
b;
    1
    ---
    105
```

We can have loops within loops, for example

```
for K:= 1:10 do for J:=1 step 2 until 7 do   ;
```

See the end of chapter 5 for other kinds of loop, which are repeated until
a certain condition holds, or while a certain condition holds. You will
not necessarily get into trouble if you use I as the index in a loop, but
it is as well to keep the habit of never using I except for the square root
of -1.

EXERCISE 4.1 Write a program to calculate $\sqrt{2}$ using the method
mentioned in exercise 2.1, using one loop for the integer ratio steps
and another for the decimal steps.

In the other common situation we call up the same calculation at
various points in the program, in each case applying this calculation in a
different context. Suppose we have a number of elementary functions in
a calculation, and wish to replace them by their Taylor expansions. We
may wish to use different arguments, to expand about different values of
these arguments, and to expand to different orders. We use a procedure,
defined once and for all in terms of dummy variables, but called up with
reference to the current values of the appropriate variables at each point
of use in the main program. Thus we have

```
procedure taylor(f,y,y0,n);
```

specified at the beginning of the program, and called up by a command such as

```
s:= taylor(sin(z),z,0,5);
```

This command means: compute the Taylor expansion of sin(z) about $z = 0$, stopping at the fifth power of z, and assign this as the value of s. In the body of the procedure taylor it is implicit that f is a function of a real variable y, and that y0 and n are respectively a real number and an integer. When we call the procedure the real variable is z, the function f is made explicit (sin(z) here) and specific values (0, 5 here) are given for y0 and n.

We have in fact already used a procedure df(y,x) in chapter 2. This is a symbolic procedure, already written in LISP and available on call using current values x, y(x). What we need now is to see how algebraic procedures, written in REDUCE, are used. Program 4.1 presents a Taylor expansion procedure in action.

PROGRAM 4.1

```
% Investigates working of a Taylor expansion procedure
procedure taylor(f,y,y0,n);
% returns the taylor expansion of f in y about y0 to
% power n
   begin scalar fact, series;
        fact:= 1;
        series:= sub(y = y0, f);
        for k:= 1:n do <<
             f:= df(f,y);
             fact:= fact*(y - y0)/k;
             series:= series + sub(y = y0, f)*fact >>;
        on div;  % to avoid large numerical denominator
        return series
   end;
TAYLOR
sx:= taylor(sin(x),x,0,7);
                 1    6      1    4     1    2
    SX:= - X*(------*X   -  -----*X  + ---*X  - 1)
                5040        120         6
ex:= taylor(exp(-x),x,0,7);
```

```
              1   7     1   6     1   5   1   4   1   3
EX:= - (-----*X -  ----*X +  ----*X - ---*X + ---*X
             5040       720       120      24      6
          1   2
        - ---*X    + X - 1)
          2
prod:= sx*ex$
let x**8 = 0;
prod;
          1   6    1   5    1   4    1   2
  - X*(-----*X  -  ---*X  +  ---*X  -  ---*X  + X - 1)
         630        90        30        3
% prod cannot be correct beyond order used in taylor
end;
```

The commands begin and end delimit a block within which the procedure makes use of local variables, which are not variables in the environment from which the procedure is called. Immediately after begin we must specify the type of local variable, for example integer or scalar (real, initialised to 0.0). The value of series is available to the main program only if we include the command return.

In order to give the series with the lowest order terms in simple form, rather than as the ratio of two large integers, the switch command div is used. In order to present the terms with the lowest order first, the switch command revpri can be used. In the main program, after getting $\sin(x)$ and $\exp(-x)$ each to order x^7, their product is truncated at x^7 by the command let x**8 = 0.

Here is another command that could be used in a Taylor expansion procedure,

```
sub(x=x0,y) + for K:=1:n sum sub(x=x0,df(y,x,K)
            * (x-x0)**K / for J:=1:K product J;
```

Notice in this that

```
for . . . sum . .
for . . .product . .
```

are expressions, unlike the instruction for .. . . do. So they can be combined like any other expressions, here for example one such expression is used as a divisor.

EXERCISE 4.2 Construct a Taylor procedure using this command. Why is this new procedure less satisfactory than the one in program 4.1? Use the switch on time, which gives the cpu time for each step in a calculation, to compare the speed of these two procedures for a moderately large *n*, like 15. The switch should be turned on at the beginning of the main program, not in the procedure.

4.2 A PROGRAM TO CALCULATE LAGRANGIAN EQUATIONS

Program 4.2 illustrates a different form of procedure specification, not using the begin . . . end block. The program also requires matrices, as well as introducing a neat way of truncating expressions formed of powers and products of several small quantities. It is based on a program given by Garrad and Quarton (1986), for the dynamics of a tower carrying two rotor blades. Their paper reports the use of REDUCE for a more elaborate wind turbine calculation, but does not print the full program. Figure 4.1 shows the essential three coordinates, which are the linear displacement of the tower *q* and the angular displacements z_1, z_2 of the blades from equilibrium positions 180° apart.

In Lagrangian dynamics the first step is to set up the potential energy *V* in terms of suitable generalised coordinates, and the kinetic energy *T* in terms of these coordinates q_j and the corresponding time derivatives dq_j/dt, or $\dot{q}_j$, denoted in the program by qjdot. Then for each *j* there is a Lagrange equation

$$d/dt\, \partial L/\partial \dot{q}_j - \partial L/\partial q_j = 0 \qquad (4.1)$$

where the Lagrangian $L = T - V$. When the coordinates are Cartesian the first term here is mass times acceleration, since *T* is proportional to mass times a sum of squares of velocities, and the second term is a force component, being a component of the gradient of the potential. In many applications the q_j and $\dot{q}_j$ are treated as small quantities, and all products and powers of order greater than two are removed from *L*. This ensures

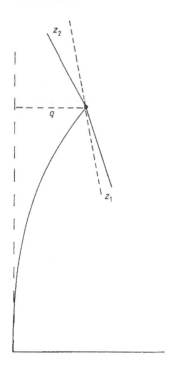

Figure 4.1 The simplified wind tower of Garrad and Quarton (1986) showing the significance of the coordinates q, z_1, and z_2. These are respectively the lateral displacement of the tower, and the lead/lag angles for each blade.

that each Lagrange equation is linear, and the problem is one of coupled small oscillations. In the present example the generalised coordinates are the displacement and angles already specified, and three Lagrange equations are required.

Since program 4.2 is the most elaborate program we have encountered so far, we present it in several pieces, starting with the two procedures used.

PROGRAM 4.2 PART 1

```
% procedure for time derivatives
procedure diffrt(a)$
begin
```

```
      return qdot*df(a,q) + psidot*df(a,psi)$
end$
% procedure for Lagrange equations
% b is a coordinate, c the corresponding velocity
procedure lag(b,c)$
begin
      dd:= df(tot,b)$
      aa:= df(tot,c)$
return omega*df(aa,psi0) + qdot*df(aa,q) + qacc*df(aa,qdot)
+ z1dot*df(aa,z1) + z1acc*df(aa,z1dot) + z2dot*df(aa,z2) +
z2acc*df(aa,z2dot) - dd$
end$
```

The first procedure sets out the time derivatives of a position vector component a in terms of velocity qdot and angular velocity psidot, and the partial derivatives of a with respect to q and psi. The second procedure obtains the left hand side of equation (4.1) for an arbitrary coordinate q, called b in the procedure, and the corresponding qdot, called c in the procedure. This procedure is specific to the coordinates of the particular problem, but applicable to each coordinate-velocity pair within that problem. It is called up three times, after the expression tot, the total kinetic energy, has been calculated. The difference from the taylor procedure in 4.1 is that there is no need to define local variables within a begin . . . end block. The procedure lag uses global variables (from the main program) such as tot, and the variables b, c which are the arguments of the procedure. No change is made in the values of those global variables which appear within the procedure, the only way the procedure affects the main program is through the return statement. When global variables appear within a procedure, we must be very careful not to change their values unless this is a desired side effect.

We now present the main program as far as the calculation of the total kinetic energy.

PROGRAM 4.2 PART 2
```
% azimuthal rotation matrix
mpsi:= mat((cos(psi), -sin(psi), 0), (sin(psi),
      cos(psi), 0), (0, 0, 1))$
```

```
% tower head displacement in inertial axes
vtow:= mat((q),(0),(0))$
% position vector of arbitrary point on blade
vr1:= mat((0),(rr),(0))$
vr2:= mpsi*vr1$
vr3:= vr2 + vtow$
% general blade velocity vector in inertial axes
vr3dot:= mat((diffrt(vr3(1,1))), (diffrt(vr3(2,1))),
(diffrt(vr3(3,1))))$
% velocity squared for arbitrary blade position
vsq:= vr3dot(1,1)**2 + vr3dot(2,1)**2 + vr3dot(3,1)**2$

let sin(psi) = sinpsi$
let cos(psi) = cospsi$
% small displacements of blades 1 and 2
psidot:= omega + z1dot$
sinpsi:= sin(psi0)*(1 - z1**2/2) + cos(psi0)*z1$
cospsi:= cos(psi0)*(1 - z1**2/2) - sin(psi0)*z1$
% velocity squared for blades 1 and 2
vsq1:= vsq$
vsq2:= sub(z1 = z2, z1dot = z2dot, sin(psi0)=
 - sin(psi0), cos(psi0) = -cos(psi0), vsq1)$
% blades are 180 degrees apart
vsqtot:= vsq1 + vsq2$
tot:= m*vsqtot/2$
let m*rr**2 = ib$  let m*rr = sb$
% blade inertia and first moment
tot:= tot$
```

The Lagrangian here is constructed in terms of a tower displacement q and two angular coordinates z_1 and z_2, one for each blade, the blades being at 180° to one another. This involves using a matrix for rotation, acting on a vector (treated as a column matrix). REDUCE allows us to define matrices, and to apply matrix multiplication for matrices of compatible dimensions, that is for $n \times m \cdot m \times k$ but not for $n \times m \cdot k \times m$, when k is not equal to m. To construct a matrix we use

```
amat:= mat((a11,a12,a13), (a21,a22,a23), (a31,a32,a33));
vmat:= mat((a1), (a2), (a3));
tmat:= mat((a1, a2, a3));
```

where each row is enclosed in ( ), and the three rows are enclosed in ( ), obtaining respectively the 3×3 matrix

$$\begin{pmatrix} a_{11} & a_{12} & a_{13} \\ a_{21} & a_{22} & a_{33} \\ a_{31} & a_{32} & a_{33} \end{pmatrix}$$

the column vector

$$\begin{pmatrix} a_1 \\ a_2 \\ a_3 \end{pmatrix}$$

and the transposed (row) vector

$$(a_1 \quad a_2 \quad a_3) .$$

However they are not presented in the output in matrix form; instead the output is, for example

```
AMAT(1,1)  := A1
AMAT(1,2)  := A12
```

and so on.

In the expression vr3dot in program 4.2 the large number of brackets is necessary because this is a 3×1 matrix, its elements are procedures, and their arguments are elements of a matrix. Once we have these matrices amat, vmat and tmat we can calculate the legitimate products such as amat*vmat, tmat*amat, tmat*vmat. We can also add matrices, multiply them by scalar quantities, transpose and invert them, and calculate the trace and the determinant of a square matrix, as in

```
diag:= trace(amat);
deta:= det(amat);
```

the characteristic polynomial of a matrix, as in

```
umat:= mat((1,0,0),(0,1,0),(0,0,1));
xpoly:= det(x* umat - amat);
```

and the inverse and transpose of a matrix,

```
invamat:= amat**(-1);
tamat:= tp(amat);
```

If we wish to introduce a new matrix name without using the `mat( )` form, or specifying the new one in terms of ones which have already been specified in this way, we can declare it

```
matrix, alpha, beta(2,3);
```

Here we get an error message if we refer to an element `alpha(i,j)`, and until we assign values to the elements of `beta` their value is 0, not `beta(i,j)`.

EXERCISE 4.3 Write a program to define the Pauli matrices

$$\sigma_x = \begin{pmatrix} 0 & 1 \\ 1 & 0 \end{pmatrix} \qquad \sigma_y = \begin{pmatrix} 0 & -i \\ i & 0 \end{pmatrix} \qquad \sigma_z = \begin{pmatrix} 1 & 0 \\ 0 & -1 \end{pmatrix}$$

and to verify the results

$$\sigma_x \cdot \sigma_y + \sigma_y \cdot \sigma_x = 0$$
$$\sigma_x^2 + \sigma_y^2 + \sigma_z^2 = 3I_2$$
$$\sigma_x \cdot \sigma_y \cdot \sigma_z = iI_2$$

where I_2 is the unit 2×2 matrix.

Returning to the Lagrangian calculation, once we have the velocities we can calculate the Lagrangian, which here is $L = T$, since no potential energy terms appear. We can now go on to the final part of the wind tower program.

PROGRAM 4.2 PART 3

```
% simplifying to get linear equations
wtlevel 2$
weight q=1, qdot=1, z1=1, z2=1, z1dot=1, z2dot=1$
let sin(psi0)**2 + cos(psi0)**2 = 1$
factor (sin(psi0), cos(psi0), omega, ib)$
```

```
tot:= tot;
   TOT:= - (2*SIN(PSIO)*OMEGA*QDOT*SB*(- Z1 + Z2) +
                                          2
2*COS(PSIO)*QDOT*SB*(Z1DOT - Z2DOT) - 2*OMEGA *IB -
                                          2      2
(2*OMEGA*IB)*(Z1DOT+Z2DOT) - IB*(Z1DOT +Z2DOT )
        2
- 2*M*QDOT )/2
% now reduced to quadratic and organised as sum and difference
clear q, qdot, z1, z1dot, z2, z2dot$
% otherwise not acceptable in lagrangian procedure
lagq:= lag(q,qdot);
LAGQ:= 2*SIN(PSIO)*OMEGA*SB*(Z1DOT-Z2DOT) + COS(PSIO)
        2
*OMEGA *SB*(Z1 - Z2) + COS(PSIO)*SB*(-Z1ACC + Z2ACC)
+ 2*M*QACC
% motion of tower coupled to relative motion of blades
lagz1:= lag(z1,z1dot);
   LAGZ1:= - (COS(PSIO)*QACC*SB - IB*Z1ACC)
lagz2:= lag(z2, z2dot);
   LAGZ2:= COS(PSIO)*QACC*SB + IB*Z2ACC
lagsum:= lagz1 + lagz2;
   LAGSUM:= IB*(Z1ACC + Z2ACC)
%  centre of mass motion uncoupled
lagdiff:= lag1 - lag2;
   LAGDIFF:= - (2*COS(PSIO)*QACC*SB + IB*(-Z1ACC
              + Z2ACC))
% relative motion of blades coupled to motion of tower
end;
```

If we had ended part 2 of this program with

```
tot:= tot;
```

the output would occupy about 15 lines (not counting lines carrying the raised exponents). The problem is unmanageable without some simplification. This we obtain by treating the problem, in part 3, as one of small oscillations. We have already set up part of this process towards the end of part 2. We use the squared sum result for sine and cosine, and the command factor, to simplify the results, and to clarify in particular the significance of $z_1 + z_2$ and $z_1 - z_2$ in this system.

The two angle variables z_1 and z_2, and the linear displacement q, are all now taken to be small. Before deriving the Lagrange equations (4.1) we remove from the Lagrangian all products and powers of z_1, z_2, q, and the corresponding velocities, beyond order 2. To do this we assign the weight level 2, and weights of 1 to z_1, z_2 etc. Only terms with weights 0, 1 and 2 are retained. This is a rather flexible trick, since we can give different variables different weights. Let us consider a single small parameter d. The weight m is assigned to a variable of order d^m. Different variables may be of different orders in d, and so have different weights. Weight level n means that all products or powers of variables giving d^{n+1} or higher are removed. Thus after determining a sum of terms which include products and powers of x and y, we can have

```
wtlevel 3;
weight x = 1, y = 2;
```

leaving only terms proportional to x, x^2, x^3, y and yx, or independent of x and y. The small parameter d is implicit in the assignment of weights.

EXERCISE 4.4 Use the Taylor expansion procedure in program 4.1 for $\sin(x)$ and $\exp(y)$, where y and x^2 are assumed to be of the same order in a small parameter. Then evaluate the product of the sine and exponential series to order 10 in this parameter.

We return once more to part 3 of the Lagrangian calculation. Since the coordinates and velocities are used after this truncation to call up the procedure lag, we need to insert

```
clear z1, z2, q, z1dot, z2dot, qdot;
```

after the truncation. In the final form of the Lagrange equations, the common motion of the two blades is uncoupled, while the motion of the tower and the relative motion of blades are coupled.

EXERCISE 4.5 Write a program to derive the Lagrange equations for a two rods AB and BC, of masses m_1 and m_2, jointed so that they are free to oscillate in the same vertical plane. One swings about its upper end A, from a block of mass m_3 which is free to slide along a horizontal line in the plane, and which is restrained by a linear restoring force of spring constant k. The other rod is suspended at B from the lower end of the first rod. Give the Lagrange equations in the small oscillation form.

4.3 A PROGRAM TO CALCULATE POISSON BRACKETS

Program 4.3 again uses a procedure, and some loops, and serves as an intermediate step between the dynamics programs of chapter 1 and the quantum mechanics programs of chapter 6. In Program 1.1 the expression GG, defined as the time derivative of G, takes the form (1.5), the Poisson bracket of G and the Hamiltonian H. The Hamilton equations (1.3) have the consequence that the Poisson bracket $\{H, G\}$, of H and a function G of coordinates and momenta, is zero when G is a constant of motion. (Notice that $\{G, H\} = - \{H, G\}$.) An interesting situation arises when there are several constants of motion in addition to H, and we can use other Poisson brackets. We do this for the atomic Kepler problem, of a central force with potential Ze^2/r. Here Z is atomic number and e is electronic charge. In the program we must not use E, which can only mean the base of natural logarithms, and therefore we denote e by EE. We use ZZ for atomic number because the coordinate z is denoted Z in our output. For any central force the angular momentum L is conserved, but for this special case there is another conserved vector, the Laplace–Lenz vector

$$W = (p \times L) - mZe^2r/r. \qquad (4.2)$$

Figure 4.2 illustrates the orientation of the four vectors, r, p, L and W.

In going from a classical problem to its quantum equivalent, results in terms of Poisson brackets go over into results in terms of commutators. Thus for the central force, the results

$$\{L, H\} = 0 \qquad [L, H] = 0$$

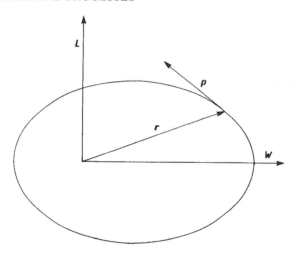

Figure 4.2 An elliptical Kepler orbit showing the orientation of the constant vectors L and W. The radius vector r and the momentum p lie in the orbital plane, the angular momentum L is perpendicular to this plane, while the Laplace–Lenz vector W lies in the orbital plane, along the major axis.

each correspond to the fact that angular momentum is conserved (we write $\{L, H\}$ as shorthand for three brackets $\{L_x, H\}$ etc.). For the Kepler force we also have

$$\{W, H\} = 0 \qquad [W, H] = 0$$

provided we take care to redefine W in a symmetrised manner in the quantum problem. Any result such as $\{L, W\} = f(L, W)$ has its analogue for the commutator $[L, W]$.

PROGRAM 4.3

```
% Poisson bracket relations in Kepler problem
procedure poisson(a,b);
        begin scalar bracket;
            bracket:= df(a,px)*df(b,x) + df(a,py)*df(b,y)
            + df(a,pz)*df(b,z) - df(a,x)*df(b,px)
            -df(a,py)*df(b,y) - df(a,z)*df(b,pz);
        return bracket
end;
```

```
POISSON
% define angular momentum and its vector product with
% linear momentum, and the Laplace--Lenz vector
operator L, LP, W$
L(1):= Y*PZ - Z*PY$
  *** L declared operator
L(2):= Z*PX - X*PZ$
L(3):= X*PY - Y*PX$
LP(1):= L(2)*PZ - L(3)*PY$
  *** LP declared operator
LP(2):= L(3)*PX - L(1)*PZ$
LP(3):= L(1)*PY - L(2)*PX$
W(1):= LP(1)/(M*ZZ*EE**2) + X/R$
  *** W declared operator
W(2):= LP(2)/(M*ZZ*EE**2) + Y/R$
W(3):= LP(3)/(M*ZZ*EE**2) + Z/R$
depend r, x, y, z$    % r is length of radial vector
let df(r,x) = x/r$ let df(r,y) = y/r$
let df(r,z) = z/r$
for J:= 1:3 do for K:= 1:J do PB(K,J):= poisson(L(J),W(K))$
  *** PB declared operator
PB1:= poisson(W(1), W(2))$ PB2:= poisson(W(1), W(3))$
PB3:= poisson(W(2), W(3))$
let x**2 + y**2 + z**2 = r**2$        on factor$
% to bring out structure of PB1, PB2, PB3
for J:=1:3 do for K:= 1:J do write PB(J,K);
0
      2          2                       2
ZZ*EE *M*Z - PX *R*Z + PX*PZ*R*X - PY *R*Z + PY*PZ*R*Y
------------------------------------------------------
                         2
                    ZZ*EE *M*R
0
      2          2                             2
ZZ*EE *M*Y - PX *R*Y + PX*PY*R*X + PY*PZ*R*Z - PZ *R*Y
------------------------------------------------------
                         2
                    ZZ*EE *M*R
```

```
0
     2          2                          2
ZZ*EE *M*X  -  PY *R*X  +  PY*PX*R*Y  -  PZ *R*X  +  PZ*PX*R*Z
-------------------------------------------------------------
                            2
                       ZZ*EE *M*R
% J = K cases must be zero, others are components of W
PB1;
       2      2       2      2
(2*ZZ*EE *M  -  PX *R  -  PY *R  -  PZ *R)*(PX*Y  -  PY*X)
---------------------------------------------------------
                        2   4  2
                       ZZ *EE *M *R
PB2;
      2      2       2      2
(2*Z*EE *M  -  PX *R  -  PY *R  -  PZ *R)*(PX*Z  -  PZ*X)
--------------------------------------------------------
                        2   4  2
                       ZZ *EE *M *R
PB3;
      2      2       2      2
(2*Z*EE *M  -  PX *R  -  PY *R  -  PZ *R)*(PY*Z  -  PZ*Y)
--------------------------------------------------------
                        2   4  2
                       ZZ *EE *M *R
% These have energy and a component of L as factors
end;
```

We come up against another problem in setting up the program to construct these Poisson brackets. How do we deal with properties of Euclidean vectors that go outside the framework of transforming a vector (row or column matrix) by multiplying it by a matrix, as in section 4.2? REDUCE was developed in the context of relativistic quantum electrodynamics, and has built in machinery for 4-vectors, but not for Euclidean vectors.

Now we can define a scalar product in matrix notation, as tp(vmat) * vmat, and we can write a procedure for a vector product of two 1 × 3 matrices as a 1 × 3 matrix (or of two 3 × 1 matrices as a 3 × 1 matrix) as we shall see in section 4.4. However this not necessarily the whole story—consider for example how a vector and a vector product transform under reflection in a plane or inversion in a point. The vector changes sign under these transformations. The vector product is an axial vector and does not change sign. Also the elements of a matrix in REDUCE can only be numbers, or variables that take numerical values, and we shall wish in chapter 6 to treat the components of a vector as operators. Reading through the output of program 4.3 we notice that various statements of the type

L DECLARED OPERATOR

appear. This is because we have introduced expressions with brackets, on the left of assignment instructions. Strictly speaking this means that these expressions, such as L(I), W(I), PB(I,J) are new operators, and require to be declared as such. We see that REDUCE takes care of this for us. In a hands-on calculation, the appearance of L(1), for example, causes the response

DECLARE L OPERATOR?

so that we can enter y and continue the calculation. We discuss operators in chapter 6, and defer discussion of the appropriate way of defining a vector product until then. In program 4.3 we set out vector products component by component.

4.4   MORE ON PROCEDURES; SCALAR AND VECTOR PRODUCTS

Suppose that we choose to treat Euclidean three-dimensional vectors as 3×1 (column) matrices. We can transform them into new vectors by multiplying by appropriate matrices, such as the rotation matrix used in section 4.2. But the scalar and vector products are not linear operations on a vector. We want a procedure that is called up in terms of the names v1 and v2 of the two vectors, not by a list of six components, and yields the scalar product as a number. (The expression tp(v1) * v2 is a 1×1 matrix.) We also want a procedure, called up in terms of the names v1 and v2, giving the vector product as a 3×1 matrix. However if we try to write a procedure vecprod (va,vb), and to use within it specific vector components va(1,2) and so on, we obtain a syntax error message. Here in program 4.4 are a procedure for the scalar product and one for the vector product.

PROGRAM 4.4

```
% testing procedures for scalar and vector products
% here the vectors are 3 x 1 matrices
matrix n1, n2, n3;
n1:= mat((1),(0),(0))$
n2:= mat((0),(1),(0))$
n3:= mat((0),(0),(1))$
procedure dot(a,b);
     begin scalar ans;
           matrix x(1,1);
           x:= tp(b)*a;
           ans:= x(1,1);   % the only element of matrix x
     return ans
     end;
DOT
procedure cross(a,b);
        mat((dot(n2,a)*dot(n3,b) - dot(n3,a)*dot(n2,b)),
            (dot(n3,a)*dot(n1,b) - dot(n1,a)*dot(n3,b)),
            (dot(n1,a)*dot(n2,b) - dot(n2,a)*dot(n1,b)));
        end;
CROSS
% cross uses both dot and the definition of the nj
% now main program tests the procedures
matrix v1(3,1), v2(3,1), vectprod(3,1), revprod(3,1);
v1:= mat((1),(2),(3));
    V1(1,1):= 1
    V1(2,1):= 2
    V1(3,1):= 3
v2:= mat((3),(-2),(1));
    V2(1,1):= 3
    V2(2,1):= - 2
    V2(3,1):= 1
```

```
scalprod:= dot(v1,v2);
   SCALPROD := 2
vectprod:= cross(v1,v2);
   VECTPROD(1,1) := 8
   VECTPROD(2,1) := 8
   VECTPROD(3,1) := - 8
zero1:= dot(vectprod,v1);
   ZERO1 := 0
zero2:= dot(vectprod,v2);
   ZERO2 := 0
% so vectprod is orthogonal to v1 and v2
revprod:= cross(v2,v1);
   REVPROD(1,1) := - 8
   REVPROD(2,1) := - 8
   REVPROD(3,1) := 8
% so revprod is - vectprod
end;
```

The procedure for vector product finds a component, $va(1,2)$ for example, by taking the scalar product of va and the matrix n2 = $((0),(1),(0))$. The set of matrices nj are defined outside the procedure, which thus uses information from the main program other than that conveyed in the arguments with which it is called. Indeed only scalar variables can be defined within the procedure. Program 4.4 uses these procedures to find $v_1 \times v_2$, and to test that it is orthogonal to v_1 and to v_2, and that $v_2 \times v_1 = -v_1 \times v_2$.

EXERCISE 4.5 Rewrite the procedures in program 4.4 so that row (1×3) matrices are used throughout.

There is much more that can be done with the vector formalism within REDUCE. REDUCE 3.4 gives access to two packages for working with vectors and curvilinear coordinates, including vector calculus; these are ORTHOVEC, developed by Eastwood (1991), and AVECTOR, developed by Harper (1989). For a concise example of the use of relativistic 4-vector machinery, see Fitch (1985), and section 17 of the REDUCE 3.3 manual.

We end this chapter with some general comments on procedures. When a return instruction is reached, the procedure is completed. However a procedure can have more than one return statement, provided these are alternatives, reached in different branches of a conditional structure. We shall deal with conditional statements in chapter 5. Here we just note that one can have, for example,

```
if a > b then return a else return b;
```

where there is no question of using both returns. On the other hand we cannot use return in a for ... statement within a procedure. A procedure need not return anything. For example one can write a procedure to clear a number of variables and turn off a number of switches. A procedure of this kind need not have any arguments, just a name,

```
procedure allclear;
```

which is called up by

```
allclear;
```

This is an example of a procedure that is intended to change the values of some global variables.

A procedure can cause a message to be displayed, for example a procedure could be used to check whether a certain expression Z contains I and to give the warning message "Z COMPLEX". A procedure can calculate several quantities a1, a2,. . and not return them; they are available to be used in the main program, since they can be recalled by a1:= a1; and so on. A procedure can call another procedure, or even itself, but cannot contain a definition of a procedure.

Summary of commands and conventions introduced in this chapter

```
for K:= 1:n do    ;    for K:= n1 step n2 until n3 do    ;
for K:= 1:n sum   ;    for K:= 1:n product   ;
procedure name(x, . .);    n:= name(y, . .)
```

local and global variables

begin scalar return only within a procedure

```
<<       >> grouping commands
on/off div   on/off   revpri
weight a:= 1, b:= 2, . . . ;   wtlevel n;
matrix bmat;
amat:= mat(( , , ),( , , ))
tp(amat)  det(amat)  trace(amat)
```

automatic operator declaration

Additions and amendments for users of 3.4

With 3.4 the user is much less likely to have to write procedures for standard processes such as Taylor expansion. In particular 3.4 brings a user package TAYLOR (R Schopf) to carry out Taylor expansions, as well as TPS (A Barnes and J Padgett), a package for manipulating series. The situation is particularly improved for vectors, with the choice of packages ORTHOVEC and AVECTOR mentioned above. 3.4 introduces ways of making the use of procedures more flexible, by removing the 'insulation' of the procedure from the main program. These involve

A using a let construction in a procedure, or

B using

```
for all x, y, let proc1(x,y) = <procedure definition>;
```

instead of

```
procedure proc1(x,y); <procedure definition>;
```

If B is used and one has

```
x:= a;
```

in the procedure, this affects x outside the procedure. If A is used without B, then

```
let x = a;
```

again affects x outside the procedure. Clearly one has to be cautious about using these variants of the procedure. Finally, there are some new operations on matrices, for example cofactor(mat,nr,nc) where mat is a matrix and nr, nc are integers specifying row and column.

```
rank(mat)
```

5

Determinants, polynomials and conditional statements

In which we examine the use of Routh–Hurwitz determinants in a problem of stability analysis, involving a generalised form of the Lorenz model, and learn about polynomials and conditional statements.

5.1 STABILITY, POLYNOMIALS AND DETERMINANTS

The stability theory of fixed points of ordinary differential equations begins with the question of stability against small displacements, using equations linearised about the fixed point. The condition for stability is expressed in terms of the characteristic polynomial $P(z)$, where z stands for a complex eigenvalue. The general solution of these linearised equations is a sum of exponentials $\exp(z_j t)$, where z_j are the roots of $P(z) = 0$. The fixed point is locally stable if and only if all roots of this polynomial are in the left hand half of the complex plane. This is a generalisation of the familar result that the fixed point (the origin) of the damped harmonic oscillator equation

$$m\mathrm{d}^2 x / \mathrm{d}t^2 + b\mathrm{d}x / \mathrm{d}t + k = 0$$

is stable provided the roots of the characteristic equation

$$z^2 + bz/m + k/m = 0$$

have negative real part, implying that a trial solution

$$x = A \exp(z_1 t) + B \exp(z_2 t)$$

decays as t increases from 0. The transition from stability to instability by way of oscillating solutions (Hopf bifurcation) involves a conjugate pair of roots crossing the imaginary axis. The standard technique for detecting this is to use the Routh–Hurwitz determinants, constructed from the coefficients of the characteristic polynomial. These determinants are required to be positive for stability. For treatments of the origin of these determinants see for example Aizerman (1963) or MacDonald (1989). The steps in linear stability analysis are

(1) Find all fixed points.

(2) Linearise about each fixed point. For each set of linear equations,

(3) Set up the Jacobian matrix (the matrix of partial derivatives) and obtain its eigenvalue equation $P(z) = 0$.

(4) Check for negative terms in the characteristic polynomial $P(z)$, which indicate stability change by way of a real root z_j changing sign. If there are no such terms,

(5) evaluate the Routh–Hurwitz determinants. We do not need to understand how the roots move about the complex plane, only whether they are all in the left hand half.

Writing the general characteristic equation as

$$P(z) = a_0 z^n + a_1 z^{(n-1)} + \ldots + a_n = 0 \tag{5.1}$$

the Routh–Hurwitz determinants are

$$D_2 = \begin{vmatrix} a_1 & a_3 \\ a_0 & a_2 \end{vmatrix} \qquad D_3 = \begin{vmatrix} a_1 & a_3 & a_5 \\ a_0 & a_2 & a_4 \\ 0 & a_1 & a_3 \end{vmatrix}$$

$$D_4 = \begin{vmatrix} a_1 & a_3 & a_5 & a_7 \\ a_0 & a_2 & a_4 & a_6 \\ 0 & a_1 & a_3 & a_5 \\ 0 & a_0 & a_2 & a_4 \end{vmatrix}$$

and so on. Any absent coefficients, such as a_5 when we use the determinant D_3 with a fourth order polynomial, are replaced by 0. It is usual to arrange that $a_0 = +1$.

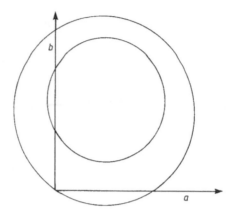

Figure 5.1   Illustrates the need for more than one Routh–Hurwitz determinant sign test. The coordinates here are two parameters a, b which determine the coefficients of an equation of order $n > 4$. Suppose we have established stability at the origin. Suppose also that on the two circles the determinant D_{n-1} is zero. Then in the annular region there must be instability, since the roots depend continuously on a, b. Without a further test we cannot tell whether stability or instability prevails within the inner circle. In the first case a pair of roots crosses the imaginary axis on the outer circle, and crosses back on the inner circle. In the second two pairs of roots successively cross the imaginary axis, one on each circle.

It is not obvious why we cannot rely on a single test for a pure imaginary pair of zeros, iy and $-iy$, to identify that the imaginary axis is crossed by a pair of roots. Such a test would be to solve the two equations

$$\mathrm{Re}(P(iy)) = 0 \qquad \mathrm{Im}(P(iy)) = 0$$

which by eliminating y indeed equates to zero a Routh–Hurwitz determinant, in fact the one of highest order in the a_i. However at order four or higher two pairs of roots may cross the imaginary axis in turn, causing ambiguity, as illustrated in figure 5. Further this determinant is also zero (Orlando's theorem) when a set of four zeros take the values $x + iy$, $x - iy$, $-x + iy$ and $-x - iy$. So for polynomials of order four and over, two or more of the determinants must be examined. Incidentally, the efficient way to obtain the determinant of highest order using REDUCE is by performing this elimination. If we denote the real and imaginary parts of $P(iy)$ by P_1, P_2, each a polynomial in y, the elimination is performed by the REDUCE command

```
resultant(P1, P2, y);
```

However the conventional form of the resultant differs from the corresponding Routh–Hurwitz determinant by a rearrangement of rows, so that one has to be cautious regarding the direction of the change from stability to instability. For example, corresponding to the Routh–Hurwitz determinant D_4, with $a_6 = a_7 = 0$, we have the resultant

$$\begin{vmatrix} a_1 & a_3 & a_5 & 0 \\ 0 & a_1 & a_3 & a_5 \\ a_0 & a_2 & a_4 & 0 \\ 0 & a_0 & a_2 & a_4 \end{vmatrix}$$

It is usual, when a Hopf bifurcation is sought, to have all the coefficients $a_n > 0$. Then, by a result known as the Lienart–Chipart theorem, only alternate Routh–Hurwitz determinants have to be evaluated. For example, for a polynomial of order 3

$$D_3 = \begin{vmatrix} a_1 & a_3 & 0 \\ a_0 & a_2 & 0 \\ 0 & a_1 & a_3 \end{vmatrix} = a_3 D_2$$

so D_3 has the same sign as D_2, and only the sign of D_2 need be tested.

5.2 AN EXTENSION OF THE LORENZ MODEL

For practice with polynomials and determinants, we use a program developed to examine the Hopf bifurcations of some generalisations of the famous nonlinear model of Lorenz (1963). This model is specified by three ordinary differential equations

$$\begin{aligned} dx/dt &= a(y - x) \\ dy/dt &= (r - w)x - y \\ dw/dt &= xy - bw. \end{aligned} \qquad (5.2)$$

(We use w here since we reserve z for the complex eigenvalue.) For equation (5.2) the fixed points of interest, for $r > 1$, are given by

$$x^2 = y^2 = b(r-1) \qquad w = (r-1). \tag{5.3}$$

The linearised equations with respect to one of these fixed points are

$$dX/dt = a(Y - X)$$
$$dY/dt = X - Y - \sqrt{(b(r-1))}W$$
$$dW/dt = \sqrt{(b(r-1))}(X + Y) - bW.$$

The characteristic equation is

$$\begin{vmatrix} z+a & -a & 0 \\ -1 & z+1 & \sqrt{(b(r-1))} \\ -\sqrt{(b(r-1))} & -\sqrt{(b(r-1))} & z+b \end{vmatrix} = 0 \tag{5.4}$$

that is

$$z^3 + (a+b+1)z^2 + b(r+a)z + 2ab(r-1) = 0$$

with all coefficients positive, and linear in r. Only one Routh–Hurwitz determinant needs to be evaluated; it is also linear in r, being

$$D_2 = a_1a_2 - a_3 = (a+b+1)b(r+a) - 2ab(r-1)$$

and leads to the stability condition

$$r < a(a+b+3)/(a-b-1).$$

There is only one critical r for loss of stability.

The interest of the Lorenz model is in the highly complicated behaviour of the trajectories when the value of r exceeds another threshold (bifurcation to chaos). This results from the nonlinearity, but also requires that the model have at least three equations. So in a sense the first (already linear) equation of the set (5.1) is there only to provide this third dimension. However one way of interpreting this equation in isolation is that the variable x follows the variable y with an exponentially decaying memory

$$x = \int_{-\infty}^{t} y(s) \exp(-a(t - s)) \, ds.$$

This introduces the concept of a mean delay for x with respect to y, $T = 1/a$. This viewpoint suggests generalising the model to include a linear chain of equations, such as

$$x = a'(u - x) = 2a(u - x)$$
$$u = a'(y - u) = 2a(y - u).$$

The time constant is doubled to keep the mean delay, which is now $T' = 2/a'$, unchanged. Now we have

$$x = \int_{-\infty}^{t} u(s)(t - s) \exp(-2a(t - s)) \, ds$$

$$u = \int_{-\infty}^{t} y(s)(t - s) \exp(-2a(t - s)) \, ds$$

and either of the fixed points has x, y, w as in (5.3) and $u = x = y$. Although the characteristic polynomials are still linear in r, the Routh–Hurwitz determinants of highest order in the coefficients are not. In fact we find that now there are two critical r values, with the fixed point stable at low and high r.

PROGRAM 5.1 PART 1

```
% Lorenz stability for clear a and b
% Characteristic determinant for original model
twodet1:= (z+1)*(z+b) + b*(r-1)$
twodet2:= b*r - z - 2*b$
threedet:= twodet1*(z+a) + twodet2*a$
order z;
% Characteristic determinants for the extended models
```

```
fourdet:= twodet1*(z+2*a)**2 + twodet2*(2*a)**2$
fivedet:= twodet1*(z+3*a)**3 + twodet2*(3*a)**3$
sixdet:= twodet1*(z+4*a)**4 + twodet2*(4*a)**4$
% These have additional delay steps but same mean delay
% fourdet is quartic in z with all coefficients positive
% Therefore only the 3 x 3 RH determinant is needed
a0:= 1$
a1:= coeffn(fourdet,z,3)$
a2:= coeffn(fourdet,z,2)$
a3:= coeffn(fourdet,z,1)$
a4:= coeffn(fourdet,z,0)$
rh1:= det(mat((a1,a3,0),(a0,a2,a4),(0,a1,a3)))/1000$
clear a1, a2, a3, a4;
% fivedet is quintic in z and we must get 2 x 2 and 4 x 4 RH
a1:= coeffn(fivedet,z,4)$
a2:= coeffn(fivedet,z,3)$
a3:= coeffn(fivedet,z,2)$
a4:= coeffn(fivedet,z,1)$
a5:= coeffn(fivedet,z,0)$
rh2:= det(mat((a1,a3),(a0,a2)))$        %  2 x 2 RH
% to get 4 X 4 RH use resultant
poly1:= x**5 - a2*x**3 + a4*x$
poly1:= poly1/x$
poly2:= a1*x**4 - a3*x**2 + a5$
let x**2 = u$
order u$
% u is -z**2 because this method sets z = ix
rh3:= resultant(poly1, poly2, u)$
% sixdet is 6th power in z, again two RH needed
clear a1, a2, a3, a4, a5;
a1:= coeffn(sixdet,z,5)$
a2:= coeffn(sixdet,z,4)$
a3:= coeffn(sixdet,z,3)$
a4:= coeffn(sixdet,z,2)$
a5:= coeffn(sixdet,z,1)$
a6:= coeffn(sixdet,z,0)$
rh4:= det(mat((a1,a3,a5),(a0,a2,a4),(0,a1,a3)))$
poly1:= -u**3 + a2*u**2 - a4*u + a6$
poly2:= a1*u**2 - a3*u + a5$
```

```
% again u is -z**2
rh5:= resultant(poly1,poly2,u)$
% now list the coefficients of powers of r in the RH
alist:= coeff(rh1,r)$
blist:= coeff(rh2,r)$
clist:= coeff(rh3,r)$
dlist:= coeff(rh4,r)$
elist:= coeff(rh5,r)$
% this is as far as we go with clear a and b
```

Program 5.1 presents the linear stability analysis for 2 and 3 equations in the linear chain. In this program we start by constructing determinants with m additional lines each containing only two entries, $-ma$ and $z+ma$. For example the extension by one equation gives the determinant

$$\begin{vmatrix} z+2a & 0 & 0 & -2a \\ -1 & z+1 & \sqrt{(b(r-1))} & 0 \\ -\sqrt{(b(r-1))} & -\sqrt{(b(r-1))} & z+b & 0 \\ 0 & -2a & 0 & z+2a \end{vmatrix}$$

This and the higher order characteristic determinants are obtained most simply by writing the original characteristic determinant (5.4) in the form $(z+a)f(z)+ag(z)$ and replacing $(z+a)$ by $(z+ma)^m$, and the other a by $(ma)^m$, rather than by using the command det with the full matrix. These changes cannot introduce any new negative coefficients, so that we can then proceed straight to the evaluation of the associated Routh–Hurwitz determinants.

We label these determinants by rh1, and so on, in the order in which we construct them. A minor point here is that our first idea here was to label them first, second, and so on. But it is not good practice to use a name, for example first, which is a special term within REDUCE (in this case a term used to identify part of a list).

For the largest Routh–Hurwitz determinant, which is the same to within sign as a resultant, we find it is faster to use the command resultant, rather than the command det, with due caution about the sign. The method used in REDUCE to obtain the resultant of two polynomials is one of polynomial division, not the evaluation of a determinant, and is thus closer to the basic process of factor-finding around which much computer algebra is built. (See for example

Davenport *et al* (1989), page 215.) Thus at each of the steps in which we evaluate a determinant it pays to be aware of the special symmetries that are appropriate.

PROGRAM 5.1 PART 2

```
on time;
% We only retain the time output when relevant
a:= 10$
b:= 8/3$
alist;
    42588    14624    704
{------, - -----,   --- }
   15       9      675
Time: 250 ms
blist;
          88
{246910,  --- }
           9
clist;
                                    147081344000    39424000
{-716277715200000,45047850720000, - ------------, - --------}
                                         3             3
Time: 1966 ms
dlist;
  8318583040000    430284800    112640
{-------------, ---------, --------}
       27           27         27
Time: 250 ms
elist;
     19230771888847298150400000000    12610877849746800640000000000
{- ---------------------------, ---------------------------,
              243                          243
   54860743023001600000000    21236465795072000000
- -----------------------, - --------------------,
             81                        243
   2952790016000000
- ----------------}
         81
```

```
Time: 1117 ms
% the sign changes indicate the number of real roots
% now we want floating point to feed to root finding program
on bigfloat;
precision 6;
6
alist;
{28392 .5, - 1624. 89, 1.042 96}
Time 316 ms
blist;
24691 0.0, 9.777 78}
clist;
{- 7.162 78 E +14, 4.504 79 E + 13, - 49027 10000 0.0, - 13141
300.0 }
Time: 666 ms
dlist;
{3.080 96 E +11, 15936 500.0, 4171. 85}
Time 300 ms
% on our machine, bigfloat uses unacceptable time for elist
end;
```

However the most important choice determining the time taken in these calculations concerns the stage at which the numerical values for a and b are introduced. We use $a = 10$, $b = 8/3$, which are the pair used most often. If we work with these values from the start the size of the integers encountered increases rapidly with the order of the characteristic determinant. We have found, in our particular implementation of REDUCE, that the third case studied in program 5.1 cannot be calculated in an acceptable time.

The end product of program 5.1 is a set of polynomials in the parameter r, for which the coefficients are listed as alist, blist, and so on. These are to be examined numerically to find critical r values. The presence of two sign changes in alist or clist implies that there cannot be a single critical r value for instability, but either two or none. In sample numerical solutions of the differential equations we find that the fixed point can be globally unstable at larger r, even though locally stable. Note also that although it seems an obvious move to convert the large numbers in these lists into floating point, we find (in our particular implementation of REDUCE) that this is very slow using `bigfloat`, and fails to convert `elist` in an acceptable time.

In general when preparing for a calculation of potentially large order, it is advisable to go step by step in case a drastic increase in running time sets in. Various readers who have worked through our examples stress that their systems can cope with this calculation, but agree that there is always a potential danger that as the order of a problem increases, the memory and CPU time requirements of computer algebra can rise exponentially.

Program 5.1 illustrates two different ways of obtaining the coefficients of successive powers in a polynomial. They can be obtained as a list, using

```
list:= coeff(poly,x);
```

and this is suitable for the final steps in part 2. But throughout part 1, where coefficients are repeatedly obtained and then fed into the next step, the `part(list,n)` form is too clumsy, and we recover the coefficients separately, using

```
am:= coeffn(poly,x,n-m);
```

(Some further operations on polynomials are introduced in chapter 7.)

EXERCISE 5.1 As the next step in the small oscillations problem treated in program 4.2, form a determinant and solve for the squared eigenfrequencies p^2, assuming solutions

$$q = Q \exp(\mathrm{i}pt + \phi).$$

EXERCISE 5.2 Write an expression in the form

```
y:= a*( ) + b*( ) + c( ) +  . . ;
```

where the brackets contain assorted functions of x. Obtain the bracketed expressions, first by using

```
abra:= coeffn(y,a,1);
bbra:= coeffn(y,b,1);
```

and so on, and then by writing

```
b:= a**2;
c:= a**3;
  . .
bralist:= coeff(y,a).
```

EXERCISE 5.3 Go back to program 3.1 and examine it to see at which point `coeffn` could be used.

EXERCISE 5.4 Set up a program, using `coeffn` and `solve`, for the following problem about the Lorenz equations (5.2), which is similar to the problems examined in chapter 1: Find the values of a, b necessary so that there is a function

$$F = (lx^2 + my^2 + z^2) \exp(nt)$$

with l, m and n not zero, and arbitrary r, satisfying $dF/dt = 0$, and determine the corresponding values of l, m, n. A general process, using REDUCE, for seeking such constants of motion has been developed by Schwartz (1985).

5.3 CONDITIONAL STATEMENTS

Suppose we want to apply the Routh–Hurwitz test for stability to an arbitrary set of linear ordinary differential equations, which we suspect may have a Hopf stability change. We need to check for sign changes between terms in the characteristic polynomial, so that we do not apply the Routh–Hurwitz test if it is irrelevant. This check can only be applied for specific parameter values.

A test like this involves a conditional statement, in which some action is taken when a Boolean variable is true, and a different action when it is false. The standard form for a conditional statement, in REDUCE as in other languages, is the `if . . then . . else . . .` statement, where the `else` clause is not always needed. Supposing that there are only three coefficients to test, we can write a procedure

```
procedure check(a,b,c);
  if a < 0 or b < 0 or c < 0
       then write "Negative coefficient, do not continue."
end;
```

and call it up, after calculating the polynomial

$$z^3 + a_1 z^2 + a_2 z + a_3$$

by

```
check(a1, a2, a3);
```

A longer version, which we give to illustrate a conditional statement in the `if . . then . . else . .` form, is

```
procedure chksgn(a,b,c);
         if a < 0 then write "do not continue";
         else if b < 0 then write "do not continue";
              else if c < 0 then write "do not continue">>
end;
```

We can use `rederr` instead of `write`. This gives the warning message and aborts the program, if a negative coefficient occurs. Or we can add a pause after the line

```
check(a1,a2,a3);
```

Whenever the two outcomes of the `if` statement require separate sequences of instructions to follow, the `if . . then . . else . .` form is needed, with one or more statements after `else`. Where more than one statement follows they are grouped by using `<<    >>`. The statements that follow `then` or `else` can include further conditional statements, as in our `chksgn` example. If a program relies on multiple choices it is very important to test with sample input that exhausts all the possibilities, so as not to be caught out by some rare but possible combination of properties.

Note that conditional statements associate to the right. Thus

```
if << test1 >> then << process1 >> else if << test2 >> then
<<process2 >> else << process 3 >>
```

is equivalent to

If test1 succeeds do process1. If it does not, apply test2. If this succeeds, do process2. If test2 also does not succeed, do process3.

and

```
if << test1 >> then if << test2 >> then <<process1 >> else
<< process2 >>
```

is equivalent to

If test1 succeeds then apply test2. If test2 succeeds (fails) do process1 (process2).

Any 'process' here implies doing something and then leaving the test sequence, to take up the following step in the calculation. Our chksgn example has the structure:
Apply test1. If it fails, apply test2. If both fail apply test3. If any of these tests succeeds do process1 and leave the test sequence, returning to the main program.

The Boolean operators that are available for use in conditional statements are = , neq (not equal), <, >, <=, >=, or, and.

The reader should note carefully where there are, and where there are not, terminators in the checking procedures. Just because there are lines without terminators all the more care is needed to get them in where they are essential. A missing terminator in a procedure can cause trouble at the beginning of the main program.

PROGRAM 5.2

```
% A collection of sign checking procedures
procedure chk1sgn(a);  % checks sign of a number
  if a < 0 then write "negative value";
end;
CHK1SGN
procedure chk2sgn(a,b);
 % checks sign of first, and if this positive, of second
    if a < 0 then write "negative value first"
    else if b < 0 then write "negative value second"
    else prod2:= a*b;
end;
CHK2SGN
procedure allneg(a,b,c);
% checks whether three signs are all negative
    if a < 0 and b < 0 and c < 0 then write "all negative"
    else write "mixed";
end;
ALLNEG
procedure oneneg(a,b,c);
% checks whether any one of three signs is negative
    if a < 0 or b < 0 or c < 0 then write "negative present"
    else prod3:= a*b*c;
end;
ONENEG
% trial numbers set out and tests applied
f:= 1$ g:= -1$ h:= 0$ k:= 2$ m:= -2$ n:= -1/2$
chk1sgn(f);
chk1sgn(g);
   negative value
chk2sgn(f,g);
   negative value second
prod2;
   PROD2
% prod2 is clear
chk2sgn(h,f);
prod2;
   0
```

```
% procedure has given prod2 a value h*f = 0
allneg(f,g,h);
   mixed
allneg(g,m,n);
   all negative
oneneg(f,g,h)
   negative present
prod3;
   PROD3
oneneg(f,h,k);
prod3;
   0
end;
```

EXERCISE 5.5 Consider the equation $ax^2 + bx + c = 0$, where any one or more of the coefficients may be zero. Write a program to give all the correct responses to $a = b = c = 0$; $a = b = 0$, $c \neq 0$; $a = 0$, $b, c \neq 0$; $a, b \neq 0$, $c = 0$; no coefficient zero.

EXERCISE 5.6 Return to the solution of the cubic equation of exercise 3.2. After identifying the condition for three real roots, use a conditional statement to help rewrite the roots so that the real or complex form of each is manifest.

Another type of instruction which involves a choice is the while . . do . . statement, a kind of modified for . . . ., as in the sequence

```
series:= 0$
J:= 0$
while (series:= series + J**2) < 100 do J:= J+1$
series;
   91
J;
   6
```

Another modified for . . . involving a choice is the repeat . . until . . statement, as in the sequence

```
series:= 0$
J:= 0$
repeat J:= J+1 until (series:= series+J**2) > 100$
series;
   140
J;
   7
```

It is up to the user to ensure that the condition tested in while . . do . . is true initially and then becomes false, or that the condition tested in repeat . . until . . is false initially and then becomes true. Otherwise the loop is either not entered or, which is worse still, it is not left.

Summary of commands and conventions introduced in this chapter

```
resultant(p1, p2, y)
coeff(poly,x)                   coeffn(poly,x,n)
if . . then . . else . .
while . . do . .                repeat . . until . .
write "remark"                  rederr "remark"
```

Boolean operators = neq < > <= >= or and

6

Non-commuting operators

In which we show how to frustrate the normal tidying up operations of REDUCE to advantage in two kinds of quantum mechanical problem.

6.1 OPERATORS

One of the features of REDUCE that accounts for its success in simplifying lengthy expressions is that it recognises, for example, that

$$x^2 y - 2xyx + yx^2 = 0$$

even when they occur interspersed with many other terms. But suppose that we wish to carry out a quantum calculation in terms of the operators x and p_x, a Cartesian coordinate and the conjugate momentum component. It is now vital not to assume that

$$xp_x - p_x x = 0$$

and remove these two terms. In fact in REDUCE we can declare x and p_x to be operators, and to be noncommuting, and we can prescribe their commutator,

$$[p_x, x] = -i\hbar.$$

Operators carry brackets, even if they have no explicit arguments, so it is as well to use $x(1)$ for x, $p(1)$ for p_x and so on. We start like this

```
operator x, p;
noncom x, p;
for all J let p(J)*x(J) = x(J)*p(J) + i*hbar;
for all J, K such that J neq K let p(J)*x(K) = x(K)*p(J);
for all J, K such that J < K let x(K)*x(J) = x(J)*x(K);
for all J, K such that J < K let p(K)*p(J) = p(J)*p(K);
```

where the notation hbar corresponds to $\hbar$ (Planck's constant divided by 2π). Thus having declared the set of six operators to be noncommutating, we have to specify zero commutator where necessary, and to adopt a convention for ordering commuting pairs. The ordering specified above ensures that a product of several momentum components and coordinates, in arbitrary order, is rearranged to have all coordinates to the left of all momentum components, and to have a standard ordering among the coordinates and among the momentum components. Any necessary cancellations will now be automatic.

Now we can go on to work out the commutators of the orbital angular momentum, the Laplace–Lenz vector, and any other expressions built up from coordinates and momenta. We have to be alert to the possibility that other normal operations of REDUCE may incidentally be frustrated. Once we decide which secondary operators we mean to calculate, we should add them to the list of declared operators at the beginning.

Thus there are four separate steps in setting up operators
(1) Declare them to be operators.
(2) In this context, declare them to be non-commuting.
(3) Add brackets so that the operators can have arguments. (We have seen in program 4.3 what happens if we write an instruction

```
a(x):=
```

where we have not declared a an operator.)
(4) Specify rules for values of operators, or of certain expressions involving the operators, their products in this case. These rules can be given either for all arguments or for specific arguments.

In the next section we also use an operator needed to calculate vector products of Euclidean vectors. This is the Levi–Civita tensor $\epsilon(j, k, m)$, which is 0 if any pair of indices are equal, and 1 or -1 when the indices are an even or odd permutation of 123. We exploit the fact that we can declare this operator to be antisymmetric. This automatically gives the zero values, and means we only need to specify $\epsilon(1, 2, 3)$, $\epsilon(2, 3, 1)$ and $\epsilon(3, 1, 2)$ to be 1, as illustrated in program 6.1.

A variety of purpose-built algebraic operators have been introduced in quantum calculations using REDUCE. Some are discussed for example by Duncan and Roskies (1986) and by Steeb (1993).

In order to introduce how non-commuting operators are used, we present three programs. Program 6.1 deals with angular momentum commutators. Program 6.2 deals with the quantum Kepler problem promised in chapter 4. Program 6.3 applies a perturbation process to a quantum oscillator. In program 6.3 the basic operators are the creation and destruction operators for phonons (shift operators for a harmonic oscillator).

6.2 COMMUTATORS FOR CONSERVED VECTORS

Program 6.1 starts from the position and momentum operators, with commutators specified as indicated above, and verifies the familiar results

$$[L(j), L(k)] = i\hbar\epsilon(j, k, m)L(m)$$

and

$$[L(j), L^2] = 0.$$

PROGRAM 6.1

```
% program for orbital angular momentum commutators
operator P, X, epsilon, L, LL, COM, commut$
antisymmetric epsilon$
noncom P, X$
% commutation relation for X(J), P(J)
for J:= 1:3 do P(J)*X(J):= X(J)*P(J) - I*HBAR$
% since P, X noncom we have to make the other pairs commute
for all J, K such that J neq K let P(J)*X(K) = X(K)*P(J)$
for all J, K such that J < K let X(K)*X(J) = X(J)*X(K)$
for all J, K such that J < K let P(K)*P(J) = P(J)*P(K)$
% to define epsilon we need only three cases
let epsilon(1,2,3) = 1$
let epsilon(2,3,1) = 1$
let epsilon(3,1,2) = 1$
% we display some cases to show effect of the antisymmetric
% declaration of epsilon
epsilon(2,1,3);
-1
```

```
epsilon(2,2,1);
0
epsilon(3,2,1);
-1
for J:= 1:3 do L(J):= for K:= 1:3 sum for M:= 1:3 sum
epsilon(J,K,M)*X(K)*P(M)$
let LL() = for J:= 1:3 sum L(J)*L(J)$
% squared angular momentum
% we now find commutators of components, and of each
% component with the square
for J:= 1:3 do COM(J):= for K:= 1:3 sum for M:= 1:3 sum
        epsilon(J,K,M)*L(K)*L(M)$
for J:= 1:3 do COM(J+3):= LL()*L(J)-L(J)*LL()$
for N:= 1:6 do write commut(N):= COM(N)
COMMUT(1):= - HBAR*I*(X(3)*P(2)-X(2)*P(3))
COMMUT(2):= HBAR*I*(X(3)*P(1)-X(1)*P(3))
COMMUT(3):= - HBAR*I*(X(2)*P(1)-X(1)*P(2))
% each commutator is proportional to the third component
COMMUT(4):= 0
COMMUT(5):= 0
COMMUT(6):= 0
% commutators with square are zero
end;
```

This program is included to show the full details of how to prescribe these operators and the ϵ operator. In the 'raw' form the products of components of L contain terms with two coordinate factors and two momentum factors, which are rearranged and tidied up by means of the let statements. The antisymmetric operator ϵ is used not only in giving components of L as a vector product of the coordinate and momentum vectors, but in giving the commutator of two components of L.

Program 6.2 evaluates the commutators of components of L and the appropriately symmetrised version of the Laplace–Lenz vector (4.2), namely

$$W = (p \times L + L \times p)/2 - Ze^2 mr/r$$

for the Kepler problem with potential $-Ze^2/r$. The commutator of a component of L and a component of W is a component of W. The commutator of two components of W, which contains terms with two coordinate factors and four momentum factors, is the product of the third component of L and the Hamiltonian,

$$H = (p_x^2 + p_y^2 + p_z^2)/2m - Ze^2/r.$$

These two vectors are orthogonal, $L \cdot W = 0$. For a state of energy E one obtains, by adding and subtracting the vectors L and $K = W/(-2mZe^2E)$, two orthogonal vectors each obeying commutation relations of the angular momentum type. They are of the same squared length. The energy E can be determined in terms of this squared length to give the standard result, in which the energy eigenvalues are proportional to $1/n^2$, with n the principal quantum number. This was the method used by Pauli to obtain the hydrogen spectrum, at the same time as the more familiar treatment by Schrödinger. The details are clearly laid out in Jordan (1986).

PROGRAM 6.2

```
operator X, P, RM, L, LP, PL, W,COM$
noncom X, P, RM$          % RM() stands for 1/r
% orthogonality and commutation patterns of vectors L and W
% components of L, angular momentum
L(1):= X(2)*P(3)-X(3)*P(2)$
L(2):= X(3)*P(1)-X(1)*P(3)$
L(3):= X(1)*P(2)-X((2)*P(1)$
% vector product of linear and angular momentum
LP(1):= L(2)*P(3)-L(3)*P(2)$
LP(2):= L(3)*P(1)-L(1)*P(3)$
LP(3):= L(1)*P(2)-L(2)*P(1)$
PL(1):= P(2)*L(3)-P(3)*L(2)$
PL(2):= P(3)*L(1)-P(1)*L(3)$
PL(3):= P(1)*L(2)-P(2)*L(1)$
% components of W, symmetrised Laplace - Lenz vector
% EE stands for charge e not energy E
W(1):= (LP(1)-PL(1))/(2*M*Z*EE**2) + X(1)*RM()$
```

```
W(2):= (LP(2)-PL(2))/(2*M*Z*EE**2) + X(2)*RM()$
W(3):= (LP(3)-PL(3))/(2*M*Z*EE**2) + X(3)*RM()$
% now provide the commutators for X,P and 1/r
for all J let P(J)*X(J) = X(J)*P(J) - I*HBAR$
for all J, K such that J neq K let P(J)*X(K) = X(K)*P(J)$
for all J, K such that J < K let X(K)*X(J) = X(J)*X(K)$
for all J, K such that J < K let P(K)*P(J) = P(J)*P(K)$
for all J let P(J)*RM() = RM()*P(J) + I*HBAR*X(J)*RM()**3$
for all J let X(J)*RM() = RM()*X(J)$
% Now give the hamiltonian
hh:= (P(1)**2 + P(2)**2 + P(3)**2)/(2*M) - Z*EE**2*RM()$
% COM(1) is scalar product, COM(2) to COM(4) are commutators
% COM(1) and COM(3) should be zero
% appropriate terms subtracted to make COM(2), COM(4) zero
COM(1):= L(1)*W(1) + L(2)*W(2) + L(3)*W(3);
   COM(1) := 0
COM(2):= L(1)*W(2)-W(2)*L(1)-I*HBAR*W(3);
   COM(2) := 0
COM(3):= L(1)*W(1)-W(1)*L(1);
   COM(3) := 0
let X(1)**2 + X(2)**2 + X(3)**2 = 1/RM()**2$
COM(4):= W(1)*W(2)-W(2)*W(1) +I*2*HBAR*HH*L(3)/(M*Z**2*EE**4);
   COM(4) := 0
end;
```

Turning to program 6.2, the only detail we wish to add is that we have to declare that $1/r$ is an operator and that it does not commute with the momentum components. Then we have to specify the commutators of $1/r$ with the momentum components. We leave the reader to improve on the somewhat repetitive form of the program, using loops as described in chapter 4.

EXERCISE 6.1 Make a more compact version of program 6.2.

6.3   PERTURBATION THEORY FOR A 1-DIMENSIONAL OSCILLATOR

Program 6.3 is adapted from one given by Sage (1988), which deals with the van Vleck method of calculating energy perturbations, for 1-dimensional and 2-dimensional harmonic oscillators. Program 6.3 deals only with the 1-dimensional case, in which the eigenstates are non-degenerate. The basic operators here are the shift operators a and ad (that is a^+) which act on an eigenstate $|n\rangle$ of the oscillator to give, to within a normalisation constant, respectively the state with energy lowered by one step or the state with energy raised by one step, $|n - 1\rangle$ or $|n + 1\rangle$. As for x, p_x before, the shift operators must be declared operators, and non-commuting, and all the necessary products must be defined. The oscillator Hamiltonian is

$$p_x^2/2m + m\omega^2 x^2/2 = (a^+ a + 1/2)h = (n + 1/2)h$$

where n is now the number operator. This Hamiltonian is perturbed by a small x^4 term, which in terms of the operators a and ad is (ad + a)^4 /4. This term couples any eigenstate $|n\rangle$ to the four eigenstates $|n - 4\rangle$, $|n - 2\rangle$, $|n + 2\rangle$ and $|n + 4\rangle$. The perturbation series for the energy is only asymptotically convergent for this perturbation. This perturbation method makes successive unitary transformations of the total Hamiltonian

$$H' = U^+ HU = \exp(-ipS)H\exp(ipS)$$

giving

$$H' = \sum_{n=0} p^n i^n [H, S]_n/n!.$$

Here

$$[H, S]_0 = H \qquad [H, S]_n = [[H, S]_{n-1}, S].$$

By this process the off-diagonal terms can be made to cancel to order 1, 2, ... in p, leaving an effective diagonal energy term.

PROGRAM 6.3

```
% 1 dimensional oscillator with quartic term
% van Vleck perturbation method
operator a, ad, h, hd, n, s$
let a()*ad() = ad()*a() + 1$
let ad()*a() = n()$  % number operator
let n()*a() = a()*n() - a()$
let n()*ad() = ad()*n() + ad()$  % shift operators

h(0,0):= 2*n()+1$   % scaled Hamiltonian of harmonic oscillator
% now calculate matrix elements of perturbing term
h(1,4):= ad()**4/4$
h(1,2):= 1/4*for k:= 0:3 sum ad()**k*a()*ad()**(3-k)$
h(1,-2):= 1/4*for k:= 0:3 sum a()**k*ad()*a()**(3-k)$
h(1,-4):= a()**4/4$
h(1,0):= (ad()+a())**4/4 - for k:= -4 step 2 until 4 sum
         k*h(1,k)/k$
for k:= -4 step 2 until 4 do s(1,k):= k*i*h(1,k)/2/k**2$
comment  In the last two lines expressions beginning with k
factor get zero value from this k, even if k or k**2 appears
in denominator$
for k:= - 6 step 2 until 6 do h(2,k):= i/2*
     for m:= max(k-4,-4) step 2 until min(k+4,4)
         sum h(1,k-m)*s(1,m) - s(1,m)*h(1,k-m)$
for n:= 2:3 do
     for k:= (-4-2*n) step 2 until (4+2*n) do
         h(n+1,k):= i*n/(n+1)/(n-1)*
             for m:= max(k-4,-4) step 2 until min(k+4,4)
                 sum h(n,k-m)*s(1,m) - s(1,m)*h(n,k-m)$
% this triple sum is by far the slowest step
for k:= - 6 step 2 until 6 do s(2,k):= i*k*h(2,k)/2/k**2$
for n:= 0: 3 do hd(n):= h(n,0)$
hd(4):= h(4,0) + i/2*
         for k:= -6 step 2 until 6 sum h(2,k)*s(2,-k) - s(2,-k)
         *h(2,k)$
for n:= 0:4 do write hd(n):= hd(n);
HD(0) := 2*N() + 1
```

$$HD(1) := \frac{3*(2*N()^2 + 2*N() + 1)}{4}$$

$$HD(2) := - \frac{34*N()^3 + 51N()^2 + 59*N() + 21}{16}$$

$$HD(3) := \frac{145*N()^4 + 290*N()^3 + 554*N()^2 + 409*N() + 132}{32}$$

$$HD(4) := - (12882*N()^5 + 32205*N()^4 + 89564*N()^3 + 102141*N()^2 + 73942*N() + 21021) / 1024$$

end;

Turning now to program 6.3 in detail, we point out several features. The process is carried to second order, giving corrections to the energy up to fourth order. The greatest part of the time taken is in a triple loop, with a for . . do loop enclosing another such loop enclosing a for . . sum expression. The number operator n is not defined explicitly but by giving its commutators with a and ad. The off-diagonal energy matrix elements h and diagonal energy increments hd must be declared operators, and so must s. The notation a^b is an alternative to a**b, but is not available in all implementations of REDUCE. The expressions k*h(1,k)/k and k*h(1,k)/k² are used so that a sum over positive and negative k values can formally include $k = 0$. This works because these expressions are evaluated from the left; the first k gives zero and the last factor is not seen. This also means that the line assigning a value to h(1,0) is not self-referring, since the program never looks for h(1,0) in evaluating the left hand side. Finally, the use of n as an index does not interfere with the use of n() as an operator.

EXERCISE 6.2 Use the operators a, ad and their analogues for two other directions to construct components of orbital angular momentum, and verify the commutators of these components.

Summary of commands and conventions introduced in this chapter

```
operator  o,p,
noncom o,p,
antisymmetric o
o():=        let o() =       o(1,2):=    let o(1,2) =
for all J let o(J) =
for all J,K such that K < J let o(J,K) =
for all J,K such that K neq J let o(J,K) =
```

power symbol ⌃ in place of ** (not always available) product evaluates from left

7

Harmonic balance

In which we investigate an approximation method for nonlinear oscillations, and look at arrays, and more ways of manipulating polynomials and other rational functions. We also examine the systematic identification of parts of a REDUCE output expression, so that it can be put in compact form.

7.1 THE HARMONIC BALANCE METHOD FOR NONLINEAR OSCILLATORS

In this chapter we build on the knowledge of REDUCE surveyed so far, in order to assess what are the likely gains from using symbolic computation in a particular context, and to lay out at least the first steps in a systematic application of REDUCE. We select for study a method of successive approximations for nonlinear oscillators, known as harmonic balance. In this method the lowest order approximation, and in simpler cases the lowest two, are readily managed without the aid of symbolic computation. However higher orders become excessively complicated. Also comparison of several different oscillators, or of several variants of the basic method, is extremely tedious even at second order. So the method provides a realistic case where there are considerable potential advantages in symbolic computation. Although apparently a rather special method, harmonic balance has strong affinities with other important numerical approximation methods, as we outline at the end of section 7.3. It has a long and complicated history, especially if one includes the related describing function method, but is currently of interest in a variety of physical applications. For example Genesio and Tesi (1992) have used it in a heuristic method for predicting the parameter ranges where chaos may occur.

We present two treatments of the method, one of which (see section 7.2) is rather long winded, and is primarily intended to illustrate the use of arrays in REDUCE, rather than as a convenient basis for harmonic balance calculations in general. These arrays here implement a table of coefficients relating two basis sets for polynomials, namely powers and Chebychev polynomials. The other method (see section 7.3) is more direct, and is the one which we continue to use throughout this chapter. It is an application of the `for all . . let . .` commands which transform between cosine products and cosine sums. We apply this method to a variety of problems, both for different oscillators and for different variants of the harmonic balance approximation.

First we have to explain the harmonic balance method. This goes back to early work by Duffing on anharmonic oscillators, and it is described in many texts, such as Mickens (1981). Suppose we wish to investigate periodic solutions of the nonlinear equation

$$L(x) = F(x, \mathrm{d}x/\mathrm{d}t, \ldots) \tag{7.1}$$

in which L is a linear operator, containing derivatives up to $\mathrm{d}x^n/\mathrm{d}t^n$, and F is a nonlinear function of x and of the derivatives up to $\mathrm{d}x^{n-1}/\mathrm{d}t^{n-1}$. Here we shall take L to be $\mathrm{d}^2x/\mathrm{d}t^2 + kx$, and F to be a function of x and $\mathrm{d}x/\mathrm{d}t$ only, although the method to be described can be applied to higher order equations. Equations such as (7.1) are frequently used to model oscillating systems in mechanics, circuit theory and biomathematics. They are classified as conservative, when there is a first integral, or dissipative when there is not. Conservative oscillators have an infinite number of periodic solutions, with the initial conditions determining which solution applies. In the second order case considered, the solutions can be visualised as nested closed loops in phase space, in which $\mathrm{d}x/\mathrm{d}t$ is displayed against x. On the other hand useful second order dissipative oscillators have one periodic solution (a stable limit cycle) which is approached asymptotically from any initial conditions. So in the phase space there are inward and outward spirals approaching a single closed loop (see figure 7.1).

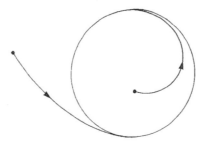

Figure 7.1 A dissipative system with a limit cycle, taken for simplicity to be the unit circle. An example with this limit circle is

$$dx/dt = y + x(1 - x^2 - y^2)$$
$$dy/dt = -x + y(1 - x^2 - y^2).$$

The particular initial conditions are immaterial, affecting only the final phase of the circular motion. So when using harmonic balance we are not free to assign an arbitrary A, with p adjusting to match it.

Suppose also that we have no reason to treat F as a small nonlinear term, so that we cannot use perturbation theory to get approximate solutions for equation (7.1). One possible tactic is to use the method of harmonic balance. In its simplest form, this method takes a trial solution of harmonic form,

$$x_0(t) = A \cos(pt)$$

evaluates

$$F(A \cos(pt), -pA \sin(pt), \ldots)$$

and truncates this function to remove higher harmonics. One then obtains two algebraic equations by setting coefficients of $\cos(pt)$ and of $\sin(pt)$ equal to zero, and solves these for the trial amplitude A and the frequency p.

For a conservative system, with dx/dt not appearing (or as we shall see in some cases with a term $(dx/dt)^2$ in F), there is no $\sin(pt)$ term. As a result of this the amplitude A is arbitrary, while the frequency p depends on A. This is to be expected, since we have an infinite family of periodic solutions in the conservative case, allowing us to set A by our choice of the initial values of x and dx/dt. As a simple example consider the cubic equation

$$d^2x/dt^2 = -x^3 \tag{7.2}$$

for which the solution to this order is

$$x_0(t) = A\cos(\sqrt{3}\,At/2).$$

An obvious next step is to use the trial solution

$$x_1(t) = A\cos(pt) + B\cos(3pt).$$

This, as discussed by Mickens (1984a), yields two equations

$$A[p^2 - 3A^2/4 - 3AB/4 - 3B^2/2] = 0$$
$$-9Bp^2 + A^2/4 + 3A^2B/2 + 3B^2/4 = 0. \tag{7.3}$$

Solving the first of equations (7.3) for $p^2(A, B)$ gives

$$p^2 = 3A^2[1 + R + 2R^2]/4$$

where $R = B/A$.

Substituting in the second of equations (7.3) yields a cubic equation for R. The method requires R small, since otherwise we cannot justify cutting off the expansion. So we look for a small real solution of the cubic. This leads to an estimate $R = 0.045$. A Fourier expansion of the exact solution of equation (7.2), which is an elliptic function, gives $R = 0.04508$.

Alternatively one can assess the closeness of the second order approximation by evaluating the period T of the cyclic motion. The exact solution gives $T = 7.416/X$ for initial displacement X, and initial velocity zero. To first order $X = A$ and the harmonic balance result is $T = 4\pi/\sqrt{3}X = 7.255/X$. To second order p^2/A^2 is increased by approximately the factor $(1+R)$, but X is $A(1+R)$, giving $T = 7.415/X$. In figure 7.2 we show the three phase–plane curves (dx/dt plotted against x) for the exact solution, the approximation x_0 and the approximation x_1. These indicate that x_1 is rather a good approximation. As an alternative

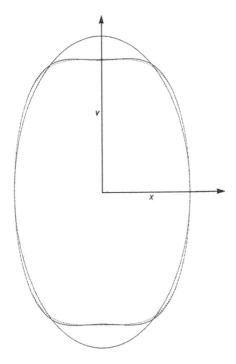

Figure 7.2 Matching the cubic oscillator orbit and the harmonic balance orbit in x, $v = dx/dt$ space. The dotted curve is a numerically integrated orbit for the cubic oscillator. The ellipse is the lowest order approximation. The other solid curve is the next approximation, which slightly exaggerates the flattened form of the true curve.

test of the quality of the approximation we can estimate the fluctuations of the energy as compared with its mean value; for the exact solution of course the energy is constant. (See exercise 7.6 for further comment on this, and for an alternative version of harmonic balance which proceeds by eliminating these fluctuations.)

Mickens (1984a) makes the following restrictions in F, for the method to be applicable.

(I) F is a polynomial; this allows expansion and truncation in an obvious manner. However various authors have applied the method to nonlinear functions of other forms. Indeed Mickens (1984b, 1989) later studied various F with singularities. One of these is a pseudo-harmonic oscillator, introduced by Mathews and Lakshmanan (1974),

$$d^2x/dt^2 + x + qx[(dx/dt)^2 + x^2]/[1 + qx^2] = 0 \qquad (7.4)$$

here the lowest order harmonic balance treatment yields the exact solution

$$x = A\cos(pt) \qquad p^2 = 1 - qA^2. \qquad (7.5)$$

(II) F is an odd function. For an even function there are difficulties which are discussed, for example, by Atadan and Huseyin (1984). These stem from the need for a constant term in the trial solution.

(III) For the conservative case, the potential has a single minimum. Otherwise one has to be clear in which neighbourhood the trial solution is to be used.

(IV) The expansion must converge; this can be established by comparing coefficients of successive terms. This should if possible be accompanied by numerical comparison with some features of the exact solution.

The significance of the method is that it provides some qualitative understanding of the way amplitude and frequency depend on parameters. Such an understanding may be laborious to extract from numerical computation with the exact solution at sampled parameter values.

7.2   CHEBYSHEV POLYNOMIAL METHOD; ARRAYS

In this section we keep the restriction to odd order polynomial form, recommended by Mickens (1984a). We shall take as our example the cubic equation with a linear term,

$$d^2x/dt^2 + kx = -mx^3. \qquad (7.6)$$

We first set out the lowest order treatment. The trial solution $x_0 = A\cos(pt)$ gives on the left hand side of (7.6)

$$A(-p^2 + k)\cos(pt)$$

and on the right hand side, before and after truncating,

$$A^3 m \cos^3(pt) = A^3 m(\cos(3pt) + 3\cos(pt))/4$$
$$= A^3 m \cos(pt)/4$$

so that the frequency satisfies

$$p^2 = 3A^2m/4 + k.$$

To the next order the left hand side is

$$A(-p^2 + k)\cos(pt) + B(-9p^2 + k)\cos(3pt)$$

while on the right we have to evaluate and truncate

$$m\{A\cos(pt) + B\cos(3pt)\}^3.$$

A possible procedure is to use

$$\cos(3pt) = (4\cos^3(pt) - \cos(pt))/3$$

within the { }, evaluate { }^{3} in powers of $\cos(pt)$, up to the ninth power, transform these powers back to sums over $\cos(mpt)$, and throw away $m > 3$. Now when $\cos\theta$ is taken as a variable, say c, then $\cos(n\theta)$ is a polynomial in c, the Chebyshev polynomial $T_n(c)$. Thus the process just set out amounts to constructing a sum of powers from two Chebyshev polynomials, raising this sum to the third power, expanding the resulting polynomial in Chebyshev polynomials, and keeping only the two of lowest odd order.

While this method is not the most concise, it gives an opportunity of introducing the topic of arrays, and some new operations on polynomials. The Chebyshev polynomials satisfy

$$T_0(c) = 1 \qquad T_1(c) = c$$
$$T_n(c) = 2cT_{n-1}(c) - T_{n-2}(c) \qquad n > 1$$

and the powers are expanded as Chebyshev polynomials by

$$c^n = (T_n(c) + U_n(c))/2^{n-1}$$

where

$$U_1(c) = 0 \qquad U_2(c) = 1$$
$$U_n(c) = 2cU_{n-1} - U_{n-2} + 2^{n-3}c^{n-2} \qquad n > 2.$$

We work from tables of the coefficients for x^m in T_n, and for T_m in x^n, from Abramowitz and Stegun (1964). These tables go up to x^{12}, which suffices for equations (7.2) or (7.6) with a $\cos(3pt)$ term in the trial solution, but not with a $\cos(5pt)$ term.

EXERCISE 7.1 If we had intended to pursue this as our standard method for all harmonic balance computations, we would have found it necessary to develop a REDUCE procedure to calculate the Chebyshev polynomials recursively. Do this.

In REDUCE, tables are implemented by arrays. An array must be declared before use, although not necessarily at the beginning of the program, as in

```
array    a(3,10), data(100), . . . ;
```

In these cases the indices run from 0 to 3, 0 to 10, 0 to 100 respectively. The elements of the array initially take the value zero. For example, if we do not assign a value to a(1,2), its value is not a(1,2), it is zero. An array can have any number of indices. The command

```
clear a;
```

means that the name a no longer stands for an array, and that all the values stored in a(i,j) are deleted. The instruction

```
clear a(1,2);
```

is only meaningful if a variable is stored in this position in the array; then that variable is cleared. This instruction can not be used to return a(1,2) to its initial value 0. The obvious use for an array, and the only one we exploit, is as a store for numerical values; a two-index array can be used as a table, a one-index array as a list of values—but not a list in the REDUCE sense.

PROGRAM 7.1

```
% Harmonic balance for oscillator with linear and cubic terms
array chp(1,1), pch(4,1);
% Chebyshev to power table
chp(1,0):= -3$          chp(1,1):= 4$
```

```
% Power to Chebyshev table
pch(1,0):= 3/4$          pch(1,1):= 1/4$
pch(2,0):= 5/8$          pch(2,1):= 5/16$
pch(3,0):= 35/64$        pch(3,1):= 21/64$
pch(4,0):= 63/128$       pch(4,1):= 21/64$
% Define left and right sides and trial solution
operator tt;
x:= a*cosp + b*cos3p$
let df(cosp,t,2) = - cosp*p**2;
let df(cos3p,t,2) = - 9*cos3p*p**2;
left:= df(x,t,2) + k*x$
% On left trial solution interpreted as cosines
xx:= a*tt(1) + b*tt(3)$
right:= - m*xx**3$
let tt(1) = cosp;
let tt(3) = chp(1,0)*cosp + chp(1,1)*cosp**3;
% On right trial solution interpreted in terms of Chebyshevs
off allfac;   order cosp;
right$
on allfac;
% Terms of successive powers have to be insulated
% from subsequent changes in lower powers
lump9:= c9*lcof(right,cosp);
                 3
   LUMP9 := - 64*B *C9*M     % cosp**9 replaced by c9
right:= reduct(right,cosp)$
lump7:= c7*lcof(right,cosp);
                 2
   LUMP7 := - 48*B *C7*M*(A - 3*B)
right:= reduct(right,cosp)$
lump5:= c5*lcof(right,cosp);
                     2            2
   LUMP5 := - 12*B*C5*M*(A  - 6*A*B + 9*B )
right:= reduct(right,cosp)$
lump3:= c3*lcof(right,cosp);
                   3   2      2      3
   LUMP3 := - C3*M*(A  - 9*A *B + 27*A*B  - 27*B )
% There is no lump1 because right side is a cube
%Now work through powers on right, converting to
%Chebyshevs, but dropping all but T1 and T3, which
%are given "left hand" names cosp, cos3p
```

```
clear right$
cc9:= pch(4,0)*cosp + pch(4,1)*cos3p$
piece9:= sub(c9 = cc9, lump9)$
% We cannot use cosp**9 directly in sub
cc7:= pch(3,0)*cosp + pch(3,1)*cos3p$
piece7:= sub(c7 = cc7, lump7)$
cc5:= pch(2,0)*cosp + pch(2,1)*cos3p$
piece5:= sub(c5 = cc5, lump5)$
cc3:= pch(1,0)*cosp + pch(2,1)*cos3p$
piece3:= sub(c3 = cc3, lump3)$
% There is no piece1 since right side is cube
right:= piece9 + piece7 + piece5 + piece3$
diff:= right - left$
% Now identify coefficients of cosp and cos3p in diff
eq1:= coeffn(diff,cosp,1);
```

$$EQ1 := -\frac{A*(3*A^2*M + 3*A*B*M + 6*B^2*M + 4*K - 4*P^2)}{4}$$

```
eq2:= coeffn(diff,cos3p,1);
```

$$EQ2 := -\frac{A^3*M + 6*A^2*B*M + 3*B^3*M + 4*B*K - 36*B*P^2}{4}$$

```
% We now begin an iterative solution. In this we first set b
% = 0 in the first equation to get an estimate of p. We use
% this approximate p in the second equation to get an estimate
% of b.
eq10:= sub(b = 0, eq1);
```

$$EQ10 := -\frac{A*(3*A^2*M + 4*K - 4*P^2)}{4}$$

```
let b:= a*r$   eq2$
sol1:= solve(eq10,p);
```

$$SOL1 := \{P = -\frac{SQRT(3*A^2*M + 4*K)}{2} ,$$

```
              2
          SQRT(3*A *M + 4*K)
   P  =   ------------------ }
              2
```
% The negative p is unphysical and is ignored
% we name the solution since we wish to substitute it in eq2
eq20:= sub(sol1,eq2);
% since p**2 appears in eq2, we need not specify second sol1
```
              2   3        2           2
          A*(3*A *M*R - 21*A *M*R + A *M - 32*K*R)
EQ20 := - ---------------------------------------
                           4
```
let r**3 = 0;   eq20;
```
              2       2
          A*(21*A *M*R - A *M + 32*K*R)
EQ20 :=   ----------------------------
                      4
```
sol2:= solve(eq20,r);
```
                2
              A *M
SOL2 := { R = ----------------- }
                2
            21*A *M + 32*K
```
% Lowest order p in sol1, lowest order b/a in sol2
% Notice how convergence is improved by linear force term
end;

Program 7.1 deals with the cubic oscillator (7.6). We declare two arrays, one containing the coefficients of powers cosp and $cosp^3$ in the expansion of the first two odd Chebyshev polynomials, and the other the coefficients of these two polynomials in expansions of each power up to $cosp^9$. The main task in program 7.1 is, after expanding x^3, to separate out the terms with decreasing powers of cosp, convert each of them into Chebyshev polynomials, and insulate them from subsequent changes caused by the conversion of lower powers. To do this we use three commands applicable when y is a polynomial in x. These are

lcof(y,x);

which gives the coefficient of the largest power of x in y,

lterm(y,x);

which gives the term with largest power of x in y, that is

```
lterm(y,x) = x**N*lcof(y,x)
```

when y is a polynomial in x of order N, and

```
reduct(y,x);
```

which is

```
y - lterm(y,x).
```

In program 7.1 we have to use both `lcof` and `lterm` because we cannot use the form

```
sub(cosp**N =    , )
```

but must replace `cosp**N` by `cN` before using `sub`. We take the calculation far enough to show that B/A is approximately $1/21$ for $k = 0$, in agreement with Mickens (1984a), and even smaller for non-zero k.

Now suppose that we wish, before beginning this 'peeling-off' process, to check the power of the expression y. We can use

```
deg(y,x);
```

which gives the highest power of x in y when y is a polynomial. Asking for `deg(y,x)` may lead to an error message if y is not a polynomial, as in

```
y:=1/x$    deg(y,x);
***** 1/x invalid as polynomial
```

but may not, as in

```
y:=exp(x)*(x^3+x+1)$
  deg(y,x);
    3
```

where the polynomial factor has been found and the exponential treated as constant. However before using these polynomial operations, it may be advisable to remove any common numerical factor; otherwise one may receive an error message '. . . not valid as polynomial'. This may depend on the particular implementation of REDUCE.

EXERCISE 7.2 Investigate what happens in program 7.1 if one separates terms with successive powers of cosp in the expression right, and uses cosp**n:=, or let cosp**n = , or match cosp**n = , to re-introduce the Chebyshev polynomials. Can one still insulate these terms from later changes?

We have not previously used the command match. To illustrate how it differs from let, take the sequence

```
let x**6 = 0;
a7:= x**7;
    A7:= 0;
match x**4 = 0;
a5:= x**5;
        5
    A5:= X
```

Thus let . . implies that all higher powers are also zero, but match . . does not.

EXERCISE 7.3 If you have used a program for solving a cubic equation, as in exercise 3.2, try transferring the results from a list to an array.

7.3   TRIGONOMETRIC EXPANSION AND CONTRACTION

An alternative method, which we apply to the cubic oscillator in program 7.2, is much simpler, especially if we wish to include $\cos(5pt)$. This method, which is likely to have occurred to the reader, is to replace all products by sums, by means of such results as

$$2\cos(3pt)\cos(pt) = \cos(4pt) + \cos(2pt),$$

until only linear terms are left, and then to remove (or just ignore) any cosine with argument greater than $3pt$. Changing the program to include

$\cos(5pt)$ in the trial solution is trivial, since there are no arrays. In fact in program 7.2, we merely ignore the higher harmonics, picking out the coefficients of $\cos(pt)$ and $\cos(3pt)$ to give eq1 and eq2.

PROGRAM 7.2

```
% Harmonic balance for oscillator with linear and cubic terms
% Uses (product of cos) to (cos of sum and difference)
x:= a*cos(p*t) + b*cos(3*p*t)$
left:= df(x,t,2) + k*x$
for all y, z let 2*cos(y)*cos(z) = cos(y+z) + cos(y-z);
for all y let 2*cos(y)**2 = cos(2*y) + 1;
% This is not implied by the previous line
right:= -m*x**3$
diff:= right - left;
% Identify coefficients of cos(pt) and cos(3pt) in diff
eq1:= coeffn(diff,cos(p*t),1)$
eq2:= coeffn(diff,cos(3*p*t),1)$
let b**2 = 0;   % we expect this coefficient to be small
eq1;
```

$$- \frac{A*(3*A^2*M + 3*A*B*M + 4*K - 4*P^2)}{4}$$

```
eq2;
```

$$- \frac{A^3*M + 6*A^2 B*M + 4*B*K - 36*B*P^2}{4}$$

```
% We now solve eq1 for p(a,b), and substitute this in
% eq2 to obtain b(a).
let p**2 = q;    % To avoid superfluous negative p solution
sol1:= solve(eq1,q);
```

$$SOL1 := \{Q = \frac{3*A^2*M + 3*A*B*M + 4*K}{4}\}$$

```
eq20:= sub(first sol1,eq2);
              3        2
           A *M  - 21*A *B*M - 32*B*K
EQ20 := - --------------------------
                        4
sol2:= solve(eq20,b);
                  3
               A *M
SOL2 := {B = ---------------}
                  2
             21*A *M + 32*K
end;
```

In developing program 7.2, we thought at one stage that there was an advantage in explicitly removing the higher harmonics. This was solely to have more compact intermediate printed results. However this led to some confusion. Since this taught us something about the care needed in the use of for all . . let . . commands, we return to this topic in section 7.4. The moral is, work with as simple a program as possible; steps that are introduced just to tidy up require just as careful checking as ones that are crucial.

Another point of detail about program 7.2 is that the transformations of $\cos^2 y$, $\sin^2 y$ have to be specified separately; they are not implied by the transformations of $\cos(y)\cos(z)$ and $\sin(y)\sin(z)$. In program 7.2 we calculate expressions for p and B/A on the assumption that B/A is small, which means that only linear equations have to be solved. The results of program 7.2 agree with those of program 7.1, and with the results quoted from Mickens (1984a).

When introducing approximations into equations, we have to consider whether the original equation will be needed later in the program. If it is, we ought to use a command of the form

```
eq10:= sub(b = 0, eq1);
```

naming the approximate equation and retaining the original equation, rather than the sequence

```
let b = 0;
eq1;
```

which transfers the original name to the approximate equation.

For a dissipative equation, in which there is a term proportional to dx/dt, both $\cos(pt)$ and $\sin(pt)$ terms appear in the lowest approximation. These are to be solved for A as well as p, since there is a unique limit cycle, not a family of periodic solutions. The van der Pol oscillator

$$d^2x/dt^2 + \epsilon dx/dt(x^2 - 1) + kx = 0 \qquad (7.7)$$

is the most famous equation of this type, used as a model first for vacuum tube oscillators and later for solid state diode oscillators, as well as for the heart beat. The first order harmonic balance result is very simple. With the trial solution again $x_0 = A\cos(pt)$, the method can be applied trivially, giving $p^2 = k$ and $A^2 = 1$, with no dependence on the parameter ϵ, which sets the scale of the nonlinearity. A similar simple result, with other values of A, characterises any equation resembling (7.7), but with the factor $(x^2 - 1)$ replaced by another even function of x. Several such equations have been used to model cases where equation (7.7) is unable to describe the observed behaviour of an experimental oscillator, and we shall use one in section 7.5. Program 7.3 applies the harmonic balance method in second order, to equation (7.7). The trial solution is now

$$x_1 = A\cos(pt) + B\cos(3pt) + C\sin(3pt)$$

and there are four simultaneous equations for p, A, B and C, obtained by setting equal to zero the coefficients of $\cos(pt)$, $\sin(pt)$, $\cos(3pt)$ and $\sin(3pt)$. The first equation is a quadratic for p, now including a term proportional to C. The second equation yields a condition, independent of p, linking A, B, and C. These results relate to the separation of p from A found at first order. The last two equations involve all four unknowns, and further progress requires numerical work, or, when the nonlinear term is small, a careful use of relative orders of smallness.

PROGRAM 7.3

```
% Harmonic balance for van der Pol oscillator
x:= a*cos(p*t) + b*cos(3*p*t) + c*sin(3*p*t)$
left:= df(x,t,2) + k*x$
right:= eps*(1-x**2)*df(x,t)$
```

```
% We retain the option to declare eps small
for all y let cos(y)**2 = (1+cos(2*y))/2;
for all y let sin(y)**2 = (1-cos(2*y))/2;
for all y, z let cos(y)*cos(z) = (cos(y+z) + cos(y-z))/2;
for all y, z let sin(y)*sin(z) = (cos(y-z) - cos(y+z))/2;
for all y, z let sin(y)*cos(z) = (sin(y+z) + sin(y-z))/2;
left$ right$   diff:= left - right$
% We have to identify four coefficients because the first
% derivative appears in the differential equation, mixing
% cosine and sine terms.
eq1:= coeffn(diff,cos(p*t),1);
```

$$EQ1 := \frac{A*(A*C*EPS*P + 4*K - 4*P^2)}{4}$$

```
eq2:= coeffn(diff,sin(p*t),1);
```

$$EQ2 := - \frac{A*EPS*P*(A^2 + A*B + 2*B^2 + 2*C^2 - 4)}{4}$$

```
eq3:= coeffn(diff,cos(3*p*t),1);
```

$$EQ3 := (6*A^2*C*EPS*P + 3*B^2*C*EPS*P + 4*B*K - 36*B*P^2$$
$$+ 3*C^3*EPS*P - 12*C*EPS*P)/4$$

```
eq4:= coeffn(diff,sin(3*p*t),1);
```

$$EQ4 := - (A^3*EPS*P + 6*A^2*B*EPS*P + 3*B^3*EPS*P + 3*B*C^2*EPS*P$$
$$- 12*B*EPS*P - 4*C*K + 36*C*P^2)/4$$

```
factor p; % Now treat b and c as small in last two equations
let b**2 = 0; let c**2 = 0; let b*c = 0;
eq3;
```

$$\frac{- 18*P^2*B + 3*P*C*EPS*(A^2 - 2) + 2*B*K}{2}$$

```
eq4;
```

$$- \frac{36*P^2*C + P*EPS*(A^3 + 6*A^2*B - 12*B) - 4*C*K}{4}$$

```
% Now insert lowest order result for a, p**2 in eq3, eq4.
% This is consistent with our use of small b, c.
let a = 2; let p = sqrt(k);
eq3;
    3*SQRT(K)*C*EPS - 8*B*K
eq4;
  - (3*SQRT(K)*B*EPS + 2*SQRT(K)*EPS + 8*C*K)
sol1:= solve(eq3,b);
                 3*SQRT(K)*C*EPS
SOL1 := {B = ----------------}
                     8*K
eq40:= sub(first sol1, eq4);
```

$$EQ40 := - \frac{16*SQRT(K)*EPS + 9*C*EPS^2 + 64*C*K}{8}$$

```
let eps**2 = 0  % small nonlinearity
eq400:= eq40;
```

$$EQ400 := - \frac{16*SQRT(K)*EPS + 64*C*K}{8}$$

```
sol2:= solve(eq400, c);
                  SQRT(K)*EPS
SOL2 := {C = - ----------------}
                     4*K
bb:= sub(first sol2, rhs first sol1);
```

$$BB := - \frac{3*EPS^2}{32*K}$$

```
% This feature, that the cos(3pt) term is of higher order in
% the (small) nonlinearity parameter than the sin(3pt) term,
% is of interest.
end;
```

At this stage we wish to set out a number of nontrivial exercises, to show how flexible the method is. They are concerned with a number of aspects of the method. First there is the question of a consistent treatment for small nonlinearity. Although lowest order harmonic balance is formally a method applicable without assuming the nonlinearity is small, in practice if we are to show that second order corrections are small we must have some small parameter in which to expand. Second there is the question of nonlinear terms which are not polynomials in the variable x, for example they could involve a rational function of x. Thirdly, can we develop the method, for a conservative oscillator, using the energy equation instead of the acceleration equation? Finally just as we can look at rational nonlinearity, what about a rational (second order) trial solution? Some of these topics are developed in MacDonald (1993).

EXERCISE 7.4 In program 7.3 we have made no use so far of eq1 and eq2. We can proceed by setting $A = 2 + d$, $p^2 = k + f$, and giving B and C the approximate values just obtained, in eq1 and eq2. Do we need to clear A, p, first? Do we get d and f of the order ϵ^2?

Alternatively, suppose we start from the original set of four equations eq1 to eq4, and expand these in the small parameter ϵ with the assumed values $A = 2 + d\epsilon^n$, $p^2 = k + f\epsilon^m$, $B = g\epsilon^r$, $C = h\epsilon^s$. It should be possible to fix both the powers m, n, r, s and the coefficients d, f, g, h, by examining the terms of lowest power in ϵ, without going through the steps in program 7.3 which take $d = 0$ and $f = 0$ in order to find g, h, r and s. The results should be consistent with the results of program 7.3 and of the first part of this exercise. Rather lengthy expansions are generated, but with care one can identify the terms of lowest order in ϵ. The results for these parameters are probably 'well known', although I have not been able to trace a reference that gives all of them.

EXERCISE 7.5 What result do you expect if the trial solution $x_1 = A\cos(pt) + B\cos(3pt)$ is used for the pseudo-harmonic oscillator (7.4), for which the exact solution is x_0? Is it necessarily the case that the solution gives $B = 0$? Notice that it is implicit in equation (7.5) that A^2 is less than $1/q$, for positive q. To help understand the various possible

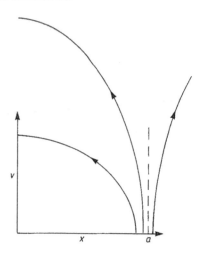

Figure 7.3   Positive quadrant of phase space for the pseudoharmonic oscillator with q positive. In bound motion $v = dx/dt$ can be arbitrarily large, while x is confined between limits $\pm X = \pm q^{-1/2}$. The period becomes very long, and the energy very large, as the point with zero velocity approaches $(q^{-1/2}, 0)$. For the case of q negative, interchange the v and x directions.

types of solution here, we show in figure 7.3 the phase space for (7.4) for positive q. Because the nonlinearity has a denominator we have in principle a choice of methods. We can obtain Fourier coefficients of the nonlinear term by integration. Can REDUCE do the necessary integrations at first order? And at second order? Or we can multiply throughout by the denominator, and apply the expansion of powers of cosines as before. (See program 7.5 for this kind of treatment.)

EXERCISE 7.6 It is easy to write the energy of the cubic oscillator (7.6) and evaluate it for the harmonic balance trial solutions. For example with the trial solution x_0 twice the energy is

$$(dx/dt)^2 + mx^4/2 = (A^4 m/32)(9 - 2\cos(2pt) + \cos(4pt))$$

where we use the result $p^2 = 3A^2 m/4$. So the amplitude of the energy fluctuations is 1/3 of the average energy. Write a program to verify this result, then try adjusting p so that the $\cos(2pt)$ term is eliminated.

Then extend this to the trial solution x_1, eliminating both $\cos(2pt)$ and $\cos(4pt)$ terms, and determining p and B/A. How different are the results from those of program 7.2 ?

EXERCISE 7.7 Mickens (1986) suggests an alternative version of harmonic balance using a rational function of cosines,

$$x_1 = A\cos(pt)/(1 + C\cos(2pt)).$$

A related method due to Prendergast (1982) has been used in astrophysics, for example by Seimenis (1989). Using this x_1, in the case of a conservative oscillator, both sides of the nonlinear equation can be put in the form of the ratio of two polynomials in $\cos(pt)$. After cross-multiplying, the process of transforming cosine products to sums will give equations to be solved for $p(A)$ and C. Try adapting program 7.2 to do this for the cubic oscillator (7.6).

When the nonlinear term is small, the alternative to harmonic balance is to develop a perturbation expansion for a periodic solution. Here the art is to adjust the timescale in order to exclude secular terms, that is to say terms which increase with time. A REDUCE program to do this for a cubic oscillator is presented, and the method expounded, in Fitch (1985).

Another related topic is the subject of a chapter in Boyd (1989). His book is concerned with boundary value problems, where the solution of a differential equation is defined over an interval and takes specific values at the ends of this interval. Besides making instructive comments on the different strategies appropriate in numerical and symbolic computation, Boyd presents one detailed REDUCE program. This is used in the case of a linear ordinary differential equation

$$L(u) = T^2 d^2u/dx^2 - u + 1 = 0$$

subject to $u(-1) = u(1) = 0$. The method involves approximating $u(x)$ up to power x^6 by

$$u_6(x) = (1 - x^2)(b_0 + b_1 x^2 + b_2 x^4 + b_3 x^6)$$

calculating the polynomial $L(u_6)$, expanding this polynomial in Legendre polynomials, which are orthogonal over the interval (-1, 1), and setting equal to zero the coefficients of the first four Legendre polynomials, to get four linear equations for the coefficients b_i.

Clearly the strategy of such calculations is close to that of harmonic balance, which applies to periodic boundary conditions. There are a wealth of such calculations, all requiring appropriate arrays to convert between powers and polynomials, or between different sets of polynomials.

7.4   INTERFERENCE OF FOR ALL ... LET ... COMMANDS

In this section we look again at the problem of insulating the effects of one command from those of another, introduced at a later stage of the program. As mentioned in section 7.3 we wished to tidy up intermediate expressions in a program resembling program 7.2, once the left and right sides of the equation were put into the form of a Fourier expansion, by setting to zero all harmonics above the third. We found that this could lead to an unexpected loss of terms. Specifically, in one calculation the sequence S1:

```
left:=     ; right:=   ;
for all y let 2*cos(y)**2 = 1 + cos(2*y);
for all y, z let 2*cos(y)*cos(z) = cos(y-z) + cos(y+z);
diff:= left - right;
```

gave an expansion of diff with terms in $\cos(pt), \cos(3pt)$ and $\cos(5pt)$. The same expansion resulted from the sequence S2:

```
for all y . . .
for all y, z . . .
left:=     ; right:=    ;
diff:= left - right;
```

However when S1 was extended to S1a by adding the lines

```
for all K such that K > 3 let cos(K*p*t) = 0;
diff;
```

one of the cos(3pt) terms was removed from `diff`. This did not happen in the corresponding extension S2a. The anomaly could be removed by replacing the line

```
for all K such that K > 3 let cos(K*p*t) = 0;
```

by

```
let cos(5*p*t) = 0;
```

or by

```
for all K such that K > 4 let cos(K*p*t) = 0;
```

Now it would clearly be dangerous to introduce the command

```
for all K such that K > 3 let cos(K*p*t) = 0
```

before the 'product to sum' commands. This is because an intermediate higher harmonic could be destroyed; for example one could not go through the process

$$4\cos(3pt)\cos(pt)\cos(pt) = 2(\cos(4pt) + \cos(2pt))\cos(pt)$$
$$= \cos(5pt) + 2\cos(3pt) + \cos(pt)$$

correctly, by applying the 'product to sum' commands, if cos(4pt) was set equal to zero in the middle of the process. We took care not to run this risk, or so we thought, and were surprised to find that separating the two types of command as in sequence S1a still led to trouble.

One way of making this calculation safe, at the cost of interrupting the flow of the calculation, is to send the harmonic expansion result to a file, converting it to a form allowed as input. The normal output form is not acceptable as input, so that all powers must be converted from raised form X^A to the form `X**A`. This is done, in the present context, by using the sequence

```
out diffstore;
off nat;
diff;
```

The default here is that the switch nat is on. The REDUCE session can then be stopped, and the rest of the calculation deferred to a fresh REDUCE session started by

```
in diffstore;
```

Then the command

```
for all K such that K > 3 let cos(K*p*t) = 0;
```

can safely be introduced, in an environment that is not aware of how the harmonic expansion in the expression diff was reached.

If we wished to isolate an intermediate output, exp:= . . . . , in a file tempstore, but continue some other aspect of the calculation, we would use the sequence

```
out tempstore;
off nat;
exp;
shut tempstore;
on nat;
```

As mentioned in 7.3, in the event the problem was resolved trivially by carrying the whole harmonic expansion forward and ignoring, rather than explicitly removing, the unwanted harmonics. However the complications just mentioned serve to emphasise the need for extreme care in using combinations of commands of the type for all . . . let . . .

7.5 HARMONIC BALANCE WITH ELLIPTIC FUNCTIONS

Bravo Yuste (1991) extends the method of harmonic balance by replacing the cosines by Jacobi elliptic functions. These can be defined as generalised cosine functions $\cos(m, pt)$ in such a way that the 'product to sum' rules used in the nonlinear part of the differential equation are unchanged. The second derivatives needed depend on the parameter m thus:

$$d^2 \cos(m, pt)/dt^2 = p^2(1 - m/2) \cos(m, pt) + p^2 m \cos(m, 3pt)$$
$$d^2 \cos(m, 3pt)/dt^2 = p^2 m \cos(m, pt) + 9p^2(1 - m/2) \cos(m, 3pt)$$
$$+ 25p^2 m \cos(m, 5pt).$$

The function $\cos(m, pt)$ reduces to $\cos(pt)$ when $m \doteq 0$. The exact solution for the cubic oscillator (7.2) is an elliptic function, $\cos(1/2, pt)$, which strongly suggests that it is worth trying this form of expansion on other power law oscillators.

PROGRAM 7.4

```
% harmonic balance for fifth power oscillator
x:= a*cos(p*t) + b*cos(3*p*t)$
% these are elliptic functions, with the same transformations
% between products and functions of sum and difference as
% cosines, but with the derivatives given explicitly in left
right:= - x**5$
left:= -p**2*(a*(1-m/2)*cos(p*t) + a*m*cos(3*p*t)/2 +
  3*b*m*cos(p*t)/2 + 9*b*(1-m/2)*cos(3*p*t))$
% m = 0 corresponds to cosines
for all y let 2*cos(y)**2 = cos(2*y) + 1;
for all y, z let 2*cos(y)*cos(z) = cos(y+z) + cos(y-z);
diff:= right - left$
% now identify coefficients of cos(p*t) and cos(3*p*t)
eq1:= coeffn(diff,cos(p*t),1)$
eq2:= coeffn(diff,cos(3*p*t),1)$
% now eliminate p**2 from these equations
let p**2 = q; eq1;
% to avoid acquiring superfluous negative p solution
         5        4        3  2        2  3           4
- (10*A  + 25*A *B + 60*A *B  + 30*A *B  + 30*A*B  + 8*A*M*Q -
      16*A*Q - 24*B*M*Q)/16
eq2;
         5        4        3  2        2  3              5
- (5*A  + 30*A *B + 30*A *B  + 60*A *B  - 8*A*M*Q + 10*B
      + 72*B*M*Q - 144*B*Q)/16
sol1:= solve(eq1,q);
SOL1 := {Q = -
          4       3        2  2        3       4
5*A*(2*A  + 5*A *B + 12*A *B  + 6*A*B  + 6*B )
------------------------------------------------- }
             8*(A*M - 2*A - 3*B*M)
```

```
% This is only apparently negative, until we fix b and m
% we cannot verify that p**2 is positive
eq20:= sub(first sol1,eq2);
              6        6        5              5
   EQ20 := - (5*(3*A *M - 2*A  - 10*A *B*M + 24*A *B -
              4 2          4 2          3 3            3 3
           45*A *B *M + 78*A *B - 108*A *B *M + 192*A *B
                 2 4          2 4        5          5
           - 84*A *B *M + 108*A *B - 52*A*B *M + 104*A*B
                  6
           - 6*B *M))/(16*(A*M - 2*A - 3*B*M))
% now fix the value of m for which b becomes zero. This
% is one way of exploiting the additional free parameter.
let b:= 0; eq20;
          5
       5*A *(3*M - 2)
    - ----------------
          16*(M - 2)
sol2:= solve(eq20, m);
              2
    SOL2 := {M = ---}
              3
% now use this m to get p**2
eq10:= sub(first sol2,eq1);
              4
           A*(15*A  - 16*Q)
   EQ10 := - ----------------
                  24
sol3:= solve(eq10,q);
                4
             15*A
    SOL3 := {Q = -----}
              16
end;
```

Program 7.4 uses this method for an x^5 oscillator,

$$d^2x/dt^2 + x^5 = 0.$$

In this program we continue to use the notation $\cos(pt)$ as shorthand for $\cos(m, pt)$. The derivatives of these new functions are given by an assignment command at the beginning,

```
left:= . . . ;
```

while the 'product to sum' rules are treated as before. The fact that a new parameter m is available in the elliptic functions allows us to use a trick which gives a single term expansion more accurate than $A\cos(pt)$. The trick is to set b equal to zero and then to use the coefficients of $\cos(pt)$ and $\cos(3pt)$ to solve for p and m. A more accurate result can be found, without going to the next trial solution, by first evaluating p and b as functions of a and m, from the coefficients eq1 and eq2, and then determining the value of m that makes the coefficient of $\cos(5pt)$ as small as possible. At each approximation step the existence of the parameter m makes possible a more accurate result than can be obtained using cosines.

7.6 AN EXAMPLE TO ILLUSTRATE THE USE OF PARTS

In this section we examine one way to abbreviate a lengthy expression by giving a name to a recurring sub-expression. We do this by identifying parts of the complete expression and replacing certain of them (those equal to the sub-expression) by the new name. Our intention here is to illustrate the use of parts in expressions rather than lists, and it happens to be convenient to use a harmonic balance example for this purpose. We use one of the equations that have been introduced to go beyond the van der Pol equation (7.7) for circuits with oscillations. Due to Walker and Connelly (1983), the equation is

$$d^2x/dt^2 + \epsilon dx/dt(x^2 - 1 + \beta)/(x^2 - 1) + x = 0. \qquad (7.8)$$

For this equation the first order harmonic balance result of Mickens (1989) is

$$p = 1 \qquad A^2 = (1 - \beta^2) \qquad (7.9)$$

where, as for the van der Pol oscillator (7.7), p and A are independent of ϵ. Program 7.5 gives preliminary calculations with this oscillator, using the method adopted by Mickens, with a rationalised version of the nonlinear equation. We once more assume small second order terms B and C, and estimate them by using first order values of p and A.

PROGRAM 7.5

```
% second order harmonic balance for Walker--Connelly
% oscillator estimates b, c from first order p,a
x:= a*cos(p*t) + b*cos(3*p*t) + c*sin(3*p*t)$
left:= (df(x,t,2) + x)*(1-x**2)$
%  We use a rationalised equation to remove denominator from
%  the right hand side. An alternative is to evaluate Fourier
%  coefficients of the original right hand side, by
%  integration.
right:= - eps*(1 - beta - x**2)*df(x,t)$    %  rationalised
for all y let cos(y)**2 = (1 + cos(2*y))/2;
for all y let sin(y)**2 = (1 - cos(2*y))/2;
for all y, z let cos(y)*cos(z) = (cos(y+z) + cos(y-z))/2;
for all y, z let sin(y)*sin(z) = (cos(y-z) - cos(y+z))/2;
for all y, z let sin(y)*sin(z) = (sin(y+z) + sin(y-z))/2;
diff:= left - right$
eq1:= coeffn(diff, cos(p*t),1)$
eq2:= coeffn(diff, sin(p*t),1)$
eq3:= coeffn(diff, cos(3*p*t),1)$
eq4:= coeffn(diff, sin(3*p*t),1)$
factor p; let c**2 = 0; let b**2 = 0; let b*c = 0;
let 1 - beta = gamma**2;
eq1$ eq2$ eq3$ eq4$
% insert first order results for a and p
let p = 1; let a = 2*gamma;
eq3;
            2                     2
16*B*GAMMA  - 8*B + 3*C*EPS*GAMMA
eq4;
                 2              2                 3
- (3*B*EPS*GAMMA  - 16*C*GAMMA  + 8*C + 2*EPS*GAMMA )
sol1:= solve({eq3,eq4},b,c);
```

```
SOL1 := {{B = -
                      2    5
               6*EPS *GAMMA
       ------------------------------------------  ,
        2    4           4          2
   9*EPS *GAMMA  + 256*GAMMA - 256*GAMMA + 64
                          3      2
               16*EPS*GAMMA *(2*GAMMA  - 1)
    C = ---------------------------------------------- }}
            2    4           4          2
        9*EPS *GAMMA  + 256*GAMMA  - 256*GAMMA  + 64
 end;
```

Program 7.5 obtains the four equations of the second order approximation, and simplifies them by assuming B and C small. Then the first order results (7.9) are substituted in the third and fourth equation to give an estimate of B and C. We note in program 7.5 that the approximate forms of B and C have the same denominator, and that this expression is liable to recur in the results when these values are substituted.

Program 7.6 takes up the calculation of program 7.5 at the stage when the estimated B and C have been substituted into the coefficient of $\cos(pt)$ to obtain the expression eq10, which is quadratic in p. The recurring expression that appears, as we see from program 7.5, is

$$9*eps**2*gamma**4+256*gamma**4-256*gamma**2+64 \qquad (7.10)$$

In program 7.6 we point to specific parts of eq10 and replace them by QUART, or -QUART as appropriate. This requires some knowledge of how parts of a complicated expression are identified without ambiguity.

PROGRAM 7.6

```
clear p, a;  % they took approximate values in 7.5
eq10:= sub(first sol1, eq1);
              3 2     2    4          4          2
EQ10 := (3*A *P *(9*EPS *GAMMA  + 256*GAMMA  - 256*GAMMA  +
            3      2    4          4          2
64) + 3*A *( - 9*EPS *GAMMA - 256*GAMMA + 256*GAMMA - 64) -
```

$$66*A^2*P^2*EPS^2*GAMMA^5 + 16*A^2*P*EPS^2*GAMMA^3*(2*GAMMA^2 - 1) +$$

$$18*A^2*EPS^2*GAMMA^5 + 4*A*P^2*(- 9*EPS^2*GAMMA^4 - 256*GAMMA^4 +$$

$$256*GAMMA^2 - 64) + 4*A*(9*EPS^2*GAMMA^4 + 256*GAMMA^4 -$$

$$256*GAMMA^2 + 64))/(4*(9*EPS^2*GAMMA^4 + 256*GAMMA^4 - 256*GAMMA^2 + 64))$$

```
part(eq10,0);
QUOTIENT
% The main operator is part0, and this is quotient
part(eq10,1)$      % The numerator
part(eq10,1,1,4);
```

$$9*EPS^2*GAMMA^4 + 256*GAMMA^4 - 256*GAMMA^2 + 64$$

```
part(eq10,1,2,3)
```

$$- 9*EPS^2*GAMMA^4 - 256*GAMMA^4 + 256*GAMMA^2 - 64$$

```
part(eq10,1,6,4);
```

$$- 9* EPS^2*GAMMA^4 - 256*GAMMA^4 + 256*GAMMA^2 - 64$$

```
part(eq10,1,7,3);
```

$$9*EPS^2*GAMMA^4 + 256*GAMMA^4 - 256*GAMMA^2 + 64$$

```
part(eq10,2,2);   % Factor in denominator
```

$$9*EPS^2*GAMMA^4 + 256*GAMMA^4 - 256*GAMMA^2 + 64$$

```
% We can now substitute all occurrences of (7.10) in eq10
eq11:= part(eq10,1,1,4):= quart$
eq12:= part(eq11,1,2,3):= - quart$
eq13:= part(eq12,1,6,4):= - quart$
eq14:= part(eq13,1,7,3):= quart$
eq15:= part(eq14,2,2):= quart;
```

$$EQ15 := (3*A^3*P^2*QUART - 3*A^3*QUART - 66*A^2*P^2*EPS^2*GAMMA^5$$

$$+ 16*A^2*P*EPS^2*GAMMA^3*(2*GAMMA^2 - 1) + 18*A^2*EPS^2*GAMMA^5$$

$$- 4*A*P \overset{2}{} *QUART + 4*A*QUART)/(4*QUART)$$

```
end;
```

The structure of the expression eq10 can be displayed in skeleton form as

$(1,1) +(1,2) +(1,3) +(1,4) +(1,5) +(1,6) +(1,7)$	$= (1)$
————————————————————	$= (0)$
$(2,1)*(2,2)$	$= (2)$

The expression QUART appears as the fourth factor in (1,1),

$$(1,1) = 3*A \overset{3}{} *P \overset{2}{} *QUART$$

as the third factor in (1,7) and as the second factor in the denominator (2), while - QUART appears as the third factor in (1,2) and as the fourth factor in (1,6).

In chapter 3 we encountered parts of a list. These are separated by commas , and are identified for example as first list or part(list,n). Here we are concerned with parts of an expression, separated by operators + * / . The operators which occur in REDUCE expressions are also identified as parts. The first step is to identify a main operator, or part 0. Given the expression exp, the main operator is identified by

```
part(exp,0);
```

For example, the expression A + B + C + D has main operator PLUS, and parts 1 to 4 are A, B, C, D respectively. The expression A + B - C also has main operator PLUS, and parts 1 to 3 are A, B, -C. The third part can be analysed further; its main part is the operator MINUS,

```
part(exp,3,0) = MINUS
part(exp,3,1) =   C
```

Analysing the expression -((A+B)/(C*D), we have

```
part(exp,0) = MINUS
part(exp,1) = (A+B)/(C*D)
part(exp,1,0) = QUOTIENT
part(exp,1,1) = A + B
part(exp,1,2) = C*D
part(exp,1,2,0) = TIMES
part(exp,1,2,2) = D
```

If we try to go too far we get an error message, for example on asking for

```
part(exp,1,2,2,1);
```

**** Expression D does not have part 1

To be sure of locating any of these subsidiary parts, for example part(exp,1,2,2), correctly, one must first look at the output expression for the part in which it is contained, in this case part(exp,1,2). The order of parts in this output may differ from that in this part of the original form of exp. So locating parts in a program requires pauses and explicit outputs.

To simplify an expression, one can replace a part by a name, in our case QUART, thus in program 7.6 we see

```
eq11:= part(eq10,1,1,1,4):= quart$
```

Here everything after the first assignment symbol means 'eq10 with the specified part replaced by QUART' . This is an expression, although it looks like a second assignment command. So we can continue with

```
solve(eq11,a);
```

for example. The effect of this part transformation is rather like that of sub( ) in

```
eq100:= sub(p = 0, eq10)
```

However, as is seen in program 7.6, one part is substituted at a time, even when the substituted name is the same. This is clumsier than sub, where we found we could substitute for a whole list of equalities. In the present case, at each successive substitution there is a chance that the order of terms is altered. So one has to take care that the correct part is identified in the next substitution.

The fact that part(exp,i,j,. .):= is an expression means that one can work with it directly, so that expressions such as

```
(part(exp1,1):= a) / (part(exp2,1,3):= b)
```

are legal, the brackets being essential here. One can also transform the main operator, as in the sequence

```
exp:= (a+b)/(c*d);
    EXP:= (A+B)/(C*D)
expo:= part(exp,1,0):=*;
    EXPO:= (A*B)/(C*D)
export:= part(expo,2,0):=+;
    EXPORT:= (A*B)/(C+D)
```

8

The REDUCE–FORTRAN interface

In which we first examine how to translate a REDUCE expression to
FORTRAN form, and then take up the question of how to construct a
program which takes advantage of the algebraic power of REDUCE and
the numerical power of FORTRAN.

8.1 TRANSFERRING RESULTS TO A NUMERICAL PROGRAM

The topic of chapter 7 leads us naturally to examine how to move
REDUCE output into a FORTRAN environment. Since there are several
possible parallel versions of harmonic balance, as suggested for example
in exercises 7.5 and 7.6, the advantage of using REDUCE is clearly
apparent in the low orders. For example it makes rapid comparisons
possible for the A-dependence of p using different methods in one
nonlinear equation. Also it allows one to survey rapidly several different
oscillators with any version of the method. However if we proceed to the
next order, in which $\cos(5pt)$ is included, for a conservative oscillator,
the resulting three simultaneous nonlinear algebraic equations get rather
heavy. With a dissipative oscillator there are already four equations
in second order, which again may not yield much qualitative insight.
REDUCE has no automatic process, analogous to solve, to handle such
sets of equations. The next steps in general have to be numerical, and it
becomes necessary to examine how we can safely transfer the output of
a REDUCE program to a numerical program. The switch

```
on fort;
```

132

ensures that any subsequent output is written in 'correct' form for FORTRAN 77. (The reason for the quotation marks will emerge shortly.)

PROGRAM 8.1

```
% converts van der Pol results to FORTRAN form
% this is a new section of an existing program
on fort; out fortout;
eq1;
      ANS=(A*(A*C*EPS*P+4.*P**2))/4.
varname eq2$  % replaces ANS by specified name
eq2;
      EQ2=-(A*EPS*P*(A**2+A*B+2.*B**2+2.*C**2-4))/4
off period;
eq3;
      EQ2=(6*A*2*C*EPS*P+3*B**2*C*EPS*P+4*B*K-36*B*P**2+3*
    . C**3*EPS*P-12*C*EPS*P)/4
% Shows what happens if new variable name not given
varname eq4;
eq4;
      EQ4=-(A**3*EPS*P+6*A**2*B*EPS*P+3*B**3*EPS*P+3*B*C**2
    . *EPS*P-12*B*EPS*P-4*C*K+36*C*P**2)/4
shut fortout;  % end belongs to the main program
```

To obtain the output shown in program 8.1, in a separate file named fortout, we have inserted in program 7.3, at the point where eqn1 to eqn4 have first been calculated for the van der Pol oscillator, the sequence of commands listed, from on fort; to shut fortout;. If we neglect to specify any variable names eq1, . . each of the expressions corresponding to eq1, . . starts in an identical manner,

        ANS :=   . . .

If we omit one variable name, here eq3, the previous one is repeated.

In the output of PROGRAM 8.1, six spaces are inserted at the left as required in FORTRAN, while in the expressions eq3 and eq4 the run-on into a second line is correctly handled, with a dot in the sixth space, and the seventh space empty. Powers and ratios are displayed as in FORTRAN. The output, with the switch fort on, gives a period after any integer, unless the integer appears as an exponent (after **). This feature is cancelled by turning off the switch period. Either form is acceptable in a FORTRAN expression. However, in numerical work it is not recommended to mix integer and floating point numbers, so if the output only contains periods in this way, the period switch is best kept on.

A complication arises if the expressions to be transferred include a labelled quantity (operator) such as X(I), with I taking integer values. With the switch period on, a period will be appended to these values. This will have to be edited out. If in our example we wish to transfer eq1, . . to a routine for solving simultaneous equations, this routine will dictate that both the equations and the unknowns be presented in labelled form. We can arrange to provide this in the REDUCE program, or alternatively these names can be changed by editing the FORTRAN file. Now editing the FORTRAN file is an inelegant way of operating, so this feature illustrates that conversion to FORTRAN needs to be thought of when planning notation for the REDUCE calculation; it is not just a matter of switching on fort.

A similar difficulty arises in the special situation in which REDUCE is used for quantum electrodynamic cross-section calculations, as described in section 17 of the 3.3 manual. In these calculations the output is given in terms of scalar products of four-vectors, denoted by an infix dot operator, thus (k1p1 + k2p2 + k3p3 − k4p4) is denoted k.p . Now a FORTRAN compiler can make nothing of this. Nor can we, within the REDUCE program, use a let command to change k.p to s(k,p), for example, which would be accepted by a FORTRAN compiler. Since we are using a symbolically coded package, which dictates the use of the dot operator, we have limited flexibility, and so it seems that these dot products have to be replaced by editing the FORTRAN output file.

Again it may be necessary for the FORTRAN lines provided by the REDUCE program to be incorporated in a program that asks for double precision. This happens for example in certain implementations of the Nag library. Then numbers such as 4. are not acceptable, and have to be replaced by symbols, with values specified in a data line in the FORTRAN program. Again this means editing the fortout file.

When a file is opened within a program, as here, using

```
out filename;
```

it can be closed again with

```
shut filename;
```

The output file fortout is only part of a FORTRAN program. It would be preceded and followed by sections written on files fort1 and fort2, so as to make up a FORTRAN program, and these merged. For example, using UNIX, this is done by the command

```
cat fort1 fortout fort2
```

8.2   THE GENTRAN PACKAGE

A user-contributed REDUCE package is available which goes beyond this simple but clumsy process of converting output into approximate FORTRAN form. This package allows one to integrate algebraic manipulations into the production of a FORTRAN program for numerical computation. The package is called GENTRAN, and is due to Barbara L Gates, RAND Corporation, Santa Monica. We shall take this in several stages, outlining only some of the possibilities. The full information, which is given in Gates (1987), issued with the REDUCE 3.3 manual, is rather lengthy. This information includes a variety of short applications, which should be worked through before embarking on an application. The GENTRAN package also offers the choice of RATFOR or C output, but we consider only FORTRAN here.

Stage 1:   Translate a REDUCE expression or command into FORTRAN, where there is a direct equivalent command. Thus there are direct equivalents to assignments := , if . . commands or

```
for J:= . .
```

commands, but there is no direct equivalent to

```
for all . . let . .
```

commands, which work in a global manner forming no part of a conventional language. For a list of translatable commands, see appendix A1 of the GENTRAN document.

Stage 2: Evaluate a REDUCE expression and then translate it. For example, if f(x) is a known expression, we can in one program line evaluate df(f,x) and translate it, using a new kind of assignment symbol :=: We can do this for an indexed set of expressions, for example to evaluate and translate the determinant of a given matrix, using ::=:

Stage 3: Provide type declarations, which are more liberally used in FORTRAN than in REDUCE. We have to exercise great care in declaring an array, because of the REDUCE convention about counting from 0. It is important to be able to declare arrays in a FORTRAN-compatible manner, if only to get round the problem with labels mentioned in section 8.1. We must declare a procedure to be a function or a subroutine. These declarations use the form, for example

```
DECLARE
    << .
       proc1 : function;
       M(4,4) : integer;
       a-h, o-z : implicit real
    >>;
```

Stage 4: Set up a template program to generate a complete FORTRAN program. The template program can have 'inactive' parts, already in FORTRAN, which are transferred to output unchanged. It has 'active' parts, which are written in REDUCE with GENTRAN operative, and are evaluated and translated before they are transferred to the output. By setting up a template program we avoid the clumsy process described in section 8.1, of merging FORTRAN-written, and REDUCE to FORTRAN translated, parts of a program by an editing process. If possible one should think through the whole calculation, assessing the potential need for numerical steps, and with GENTRAN in mind, at an early stage.

Stage 5: Provide appropriate global instructions, so that the output program can be compiled and run. For example, most compilers for FORTRAN can handle no more than 19 lines. Without needing to test the output expressions as they are formed, the switch GENTRANSEG, which is normally on, ensures that they are broken up, using suitable dummy variables, into compiler-friendly chunks. These have 800 characters maximum length, unless this is altered by

```
MAXEXPRINTLEN!* 500
```

for example. The maximum length of a line in the output can be selected by

```
FORTLINELEN!* 50
```

for example, with default 72. The output channel must also be chosen, by means of

```
GENTRANOUT filename;
```

There are many refinements possible, using a list of open files. Another language can be selected by setting

```
GENTRANLANG!*   C
```

for example, at the beginning of the program. The default setting is FORTRAN.

We hope that this brief introduction will give an idea of the useful possibilities of an integrated approach to the symbolic and numerical aspects of programming a physical problem. A further valuable extension of such an approach, involving access to the NAG library, is to be found in Davenport *et al* (1992).

Summary of commands and conventions used in chapter 8

When transferring output to a fortran file

```
on/off fort;
```

translated lines begin ANS =
varname abc makes translated lines begin ABC =
To remove/restore period after translated integers

```
off/on period;
```

When using GENTRAN

GENTRANLANG!* language name

GENTRANSEG

GENTRANOUT filename

MAXPRINTLEN!* 500

FORTLINELEN!* 50

:=:   ::=: new assignment commands

Appendix A

REDUCE 3.4

All the material in this text refers to the edition 3.3 of REDUCE. In the summer of 1991 a new edition, 3.4, was issued. There are a number of detailed changes from 3.3, for example in handling floating point arithmetic. There should be no difficulty in using this text in the context of 3.4, so long as one takes care to refer to the manual of the new edition. In notes at the end of chapters 2 and 4 we provide some indication of relevant changes from 3.3.

The main extension, and the considerable advantage of the new edition, is that it makes available many more special packages, contributed by users. REDUCE 3.3 suffers greatly in comparison with alternative computer algebra systems in that ready-made programs exist for a relatively small range of special techniques. For example most such systems will have an command such as

```
taylor(f(x),x,x0,n)
```

giving a Taylor expansion of the relevant function, to order n, about x0: with REDUCE 3.3 the user has to prepare and check a procedure, as we did in chapter 4. However, with 3.4 REDUCE a Taylor procedure can be summoned up from one of the new packages. The same applies to Laplace transforms and so on. We list the facilities available with REDUCE 3.4 briefly, with (first) authors:

ALGINT Integration for functions involving roots—J H Davenport
ARNUM Algebraic numbers—E Schrufer
AVECTOR Vector algebra—D Harper

CHANGEVAR Change of variables in differential equations—
G Uccoluk
COMPACT Condensing expressions with polynomial side relations—
A C Hearn
CVIT Dirac gamma matrices—V Ilyin
DESIR Differential equations and singularities—C Dicrescenzo
EXCALC Calculus for differential geometry—E Schrufer
FIDE Code generation for finite difference schemes—R Liska
GENTRAN Code generation in FORTRAN, RATFOR, C—B Gates
GROEBNER Multivariate polynomial ideals—H Melenk
HEPHYS High energy physics—A C Hearn
LAPLACE Laplace and inverse Laplace transform—C Kazasov
LININEQ Linear inequalities and linear programming—H Melenk
ODESOLVE Ordinary differential equations—M MacCallum
ORTHOVEC Calculus for scalar and vector quantities—J W Eastwood
PHYSOP Additional support for non-commuting quantities—M Warns
PM Algebraic pattern matcher—K McIsaac
REACTEQN Manipulation of chemical reaction systems—H Melenk
ROOTS Roots of polynomials—S L Kameny
SCOPE Optimising numerical programs—J A van Hulzen
SPDE Symmetry analysis for partial differential equations—
F Schwarz
SUM Sum and product of series—F Kako
TAYLOR Multivariate Taylor series—R Schopf
TPS Univariate Taylor series with indefinite order—A Barnes

Appendix B

An Application to Plasma Waves

Here we reprint a paper by Diver (1991), keeping to his notation, which differs in some minor ways from that used in the main text. However we follow our usual practice by distributing the five program sections through the text, rather than in an Appendix, as originally printed.

Modelling in Physics with Computer Algebra
D A Diver
Department of Physics and Astronomy
University of Glasgow

A sophisticated model for linear waves in an inhomogeneous plasma is tackled completely using the computer algebra system REDUCE. The algebra code mirrors the mathematics, and is structured in a simple and straightforward manner. In so doing, the solution technique is made obvious, and the overall philosophy of the approach is intuitive to the non-specialist computer algebra user.

B.1 INTRODUCTION

In most areas of plasma physics, the mathematical models which describe the behaviour of the plasma (or ionised gas) are extremely complicated and minimally tractable, especially when the plasma equilibrium is non-uniform. Even in the simplest descriptions, permitting inhomogeneous background parameters leads to an almost impossible algebra burden which deters a systematic theoretical investigation.

In this paper, one such simple plasma model is presented fully worked as an example of the power and utility of computer algebra in such circumstances. The paper has the twin objectives of demonstrating how plasma theory can be developed by judicious use of computer algebra techniques, and how REDUCE may be used to mirror mathematics in an intuitive manner, illuminating the method of solution, rather than the heavy mechanics of the manipulation itself. Hopefully, the latter purpose wll appeal to those non-expert programmers like myself who often are dissuaded from using algebra packages by the apparent effort needed in extracting useful results from computer runs, and who are frustrated by the lack of simple examples in the literature.

The paper starts by defining the physical model, and lays out the analytical procedures required to progress to the desired solution. Then the corresponding REDUCE structure is detailed, with discussions on style and method. Finally, the concluding section contains output from running the codes in a REDUCE environment.

B.2 THE PHYSICAL MODEL

A plasma is a fully ionised gas of electrons and ions, dominated by non-local forces resulting from the consequent electromagnetic interactions. In the simple cold plasma, the medium is treated as a perfectly conducting pressureless fluid which can support electromagnetic waves. A fuller description of basic plasma physics can be found in many texts, such as Stix (1962), or Boyd and Sanderson (1969).

Notwithstanding the physical interpretation of the model, the basic equations governing the behaviour of the plasma fluid are

$$\partial n/\partial t + \nabla \cdot (n_0 v + n v_0) = 0 \qquad (B.1)$$

$$m(\partial v/\partial t + v \cdot \nabla v_0 + v_0 \cdot \nabla v) = q(E + v_0 \times B + v \times B_0) \qquad (B.2)$$

$$\nabla \times E = -\partial B/\partial t \qquad (B.3)$$

$$\nabla \times B = \mu_0(J + \epsilon_0 \partial E/\partial t) \qquad (B.4)$$

$$J = nqv \qquad \text{(B.5)}$$

where we have considered only one species of particle for simplicity. The symbols quoted have the following physical meanings: n is the number density of particles carrying charge q having mass m and moving with velocity v; E, B and J denote respectively the electric and magnetic fields in the plasma, and the plasma current. A subscript 0 on any of these quantities denotes the equilibrium values.

Following Diver *et al* (1990) we wish to study linearised perturbations about an inhomogeneous equilibrium, in which the magnetic field B_0 has constant magnitude but varies in direction:

$$B_0 = B_0(x \cos \phi(z) + y \sin \phi(z)) \qquad \text{(B.6)}$$

We will assume a constant rotation rate throughout, i.e.

$$\phi(z) = \phi' z \qquad \phi' = \text{constant,}$$

although this is a restriction which can be relaxed. Note also that we will take a periodic time dependence of all perturbed variables, so that

$$\partial/\partial t \to -i\omega$$

when operating on any first order term. For physical reasons which are explained in Diver *et al* (1990) the equilibrium satisfies $v_0 \times B_0 = 0$, $E_0 = 0$ and $v_0 \cdot \nabla = 0$.

B.3 SOLUTION METHOD

The goal of this analysis is to describe the kinds of wave it is possible to propagate in this nonuniform plasma. Thus we must derive a wave equation, and solve it, to get this information. Since there is only one independent variable (viz. z) we expect to derive an ordinary differential equation (ODE) and it is preferable to derive a homogeneous one (i.e. only one dependent variable throughout).

The plan of attack must be as follows.

(i) Solve the vector equation (B.2) for each perturbed velocity component in terms of each perturbed electric field component (remember, since this is a linear equation, this is always possible).

(ii) Then, using these relations, express the perturbed current J_1 as a function of E_1 through equation (B.5).

(iii) Taking the curl of equation (B.3), yielding

$$\nabla \times \nabla \times E = \nabla(\nabla \cdot E) - \nabla^2 E = -\partial/\partial t \nabla \times B \qquad (B.7)$$

use equation (B.4) to substitute for $\nabla \times B$ in terms of J, so that in principle equation (B.7) contains only the perturbed electric field E_1 as the dependent variable. However, it is still a vector equation. To derive a homogeneous ODE,

(iv) Eliminate two components in favour of the third.

The final step in the plan is

(v) solve the resulting ODE.

As a scheme, this is not too difficult to define. However, the algebraic manipulation involved is very tedious, and extremely heavy. It is at this stage that the theorist may be tempted to make simplifications to ease the burden of the mathematics, thus compromising the generality and accuracy of any subsequent analysis.

However, the next section illustrates how this solution plan may be implemented without regard to the overheads, by incorporating the structure in basic REDUCE commands.

B.4 HARDWARE AND SOFTWARE

Before discussing the construction of the REDUCE code, it is important to describe the computing environment used in this work. The work was performed on a High Level Hardware ORION 1/05, running OTS as an operating system (a direct port of UNIX 4.2), and the software was REDUCE 3.3, running Kyoto Common Lisp (KCL).

B.5 CONSTRUCTING THE REDUCE CODE

Having constructed the best route to a solution, the next major task is to construct this path in REDUCE instructions.

B.5.1 Notation

Since it is not possible to use the same notation in both the text of this document and the program listings, we must relate the code variables to the model notation. As a general rule, any variable name ending in x, y or z refers to the x, y or z component of that vector quantity. The equilibrium variables are identified in REDUCE by having the affix 0, with the perturbed unknown quantities sharing the same names as in the text, but devoid of the subscript 1. Thus Ex is the x-component of E_1, and so on.

B.5.2 Assigning the variables

The first stage must be to define the basic parameters of the problem, such as the direction of the equilibrium fields, the functional dependence of the perturbed quantities, and so on. Thus we must examine the equilibrium, given by setting $\partial/\partial t$ of any variable to zero in the set (B.1)–(B.5). This allows us to consider a time-independent equilibrium, which is force-free if we also set $E_0 = 0$, and then demand $v_0 \times B_0 = 0$. Note that $v_0 \neq 0$, from (B.4), (B.5) and (B.6). These aspects of the model plasma are defined by the REDUCE program SETUP, quoted below

```
%                   Program SETUP
Comment  this file gives the equilibrium magnetic fields and
velocities for the cold plasma model with rotating magnetic
field. It is intended that this file is the setup routine for
running stage1 during a REDUCE session.$

Pause;
Depend cosp,z$ Depend sinp,z$ % direction cosines function(z)
B0x:=B0*cosp$    B0y:= B0*sinp$    B0z:= 0$
ux:= u0*cosp$    uy:= u0*sinp$    uz:= 0$
Depend Ex,z$     Depend Ey,z$      Depend Ez,z$
Depend Bx,z$     Depend By,z$      Depend Bz,z$
Depend vx,z$     Depend vy,z$      Depend vz,z$
Df(cosp,z):= -sinp*dphi$    Df(sinp,z):= cosp*dphi$
let cosp^2 + sinp^2 = 1$
Comment  at this stage, all equilibrium quantities should
be defined, and all quantities should have the correct
dependencies. Continue by calling up stage1.$
end;
```

In order to ease the notation, we have denoted the equilibrium velocity by u. Thus the magnitude of the equilibrium velocity is u_0, and that of the equilibrium field is B_0. All variables are perturbed quantities unless they possess the affix 0. Thus all the first order variables have been made functions of z, by virtue of the Depend statements involving their x, y and z components.

B.5.3 Coding the model equations

Having set up the basic problem, and assigned the appropriate variable dependencies, we must now tackle the full model differential equations. Clearly, (B.2) is the crucial equation to solve in order to express the perturbed velocity in terms of the perturbed electric field. Thus we must write the REDUCE equivalent of (B.2), component by component. This is done after the first Pause statement in program STAGE1, listed below. Notice that the advective derivative has been expanded fully, as has the vector cross-product.

Note that the perturbed magnetic field B_1 appears on the right hand side of (B.2) after linearisation. It can be eliminated in favour of E_1 via equation (B.3). This is done, component by component, after the second Pause. Now REDUCE can substitute for B_1 retrospectively, enabling the set eqn1, eqn2 and eqn3 to be solved for v_1 in terms of E_1, and constructing a list of the answers in velist. Thus the first stage in the solution technique has been achieved with minimal effort.

Now that we have v_1 in terms of E_1, we want to construct the perturbed current J_1 using equation (B.4). However, when we linearise (B.4), we see that the perturbed number density enters the calculation for the first time. This must be eliminated in order to progress to a vector equation in E_1 alone. This is achieved using the continuity equation (B.1), which when linearised yields n_1 as a function of v_1. Note that this gives n_1 directly in terms of the components of E_1, since the REDUCE environment can already eliminate v_1 in favour of E_1. Hence, we need do nothing other than write out equation (B.1) in REDUCE, and solve it for n_1. This is the task undertaken by the program STAGE2.

```
%                    Program STAGE1
Comment  this is the first step in constructing the equations
governing a cold plasma with a rotating magnetic field. First
we write down the equations of motion governing a species s
and solve for the perturbed velocity vs in terms of the
perturbed electric field E, with the equilibrium magnetic
field B0 and equilibrium velocity us.$
Pause;
eqn1:= -i*w*m*vx + m*vz*df(ux,z) - q*(Ex + vy*B0z - vz*B0y +
          uy*Bz - uz*By)$
eqn2:= -i*w*m*vy + m*vz*df(uy,z) - q*(Ey + vz*B0x - vx*B0z +
          uz*Bx - ux*Bz)$
eqn3:= -i*w*m*vz - q*(Ez + vx*B0y - vy*B0x + ux*By - uy*Bx)$
Pause;
Comment now solve for B in terms of E using dB/dt = - curlE;
Bx:= (df(Ez,y) - df(Ey,z))/(i*w)$
By:= (df(Ex,z) - df(Ez,x))/(i*w)$
Bz:= (df(Ey,x) - df(Ex,y))/(i*w)$

Comment  now solve for v in terms of E;
velist:= solve({eqn1,eqn2,eqn3},{vx,vy,vz})$
Pause;
vx:= rhs first first velist$
vy:= rhs second first velist$
vz:= rhs third first velist$
let q*B0/m = omc$        % defining the cyclotron frequency
Comment now we have v as a function of E and E'. Next we must
calculate the current in terms of E and E'. To do this, we
must take into account the number density perturbations.
Comment this ends stage1. To continue with the calculation,
load stage2;
end;
```

Now we can proceed directly to expressing the perturbed current J_1 in terms of E_1 using equation (B.4), and this is done after the first Pause in STAGE2, thus completing stage (iii) in section B.3.

The construction of the vector set of ODEs, stage (iv), involving only the components of E_1 as the unknowns, depends on using equation (B.7). Again, it is sufficient to write out the components of (B.7) in order to effect the required substitutions and eliminations automatically. This is done after the second Pause in STAGE2.

The last instructions in STAGE2 allow the z component of E_1 to be eliminated in the set of equations in favour of the other two components. This can be done directly, because no derivative of E_{1_z} occurs in the z component of (B.7). Hence we now have two remaining coupled ODEs in two unknowns,

---

```
                    Program STAGE2
Comment this is the next stage in constructing the model
equations for the cold plasma with rotating magnetic field.
The perturbed current is expressed in terms of the perturbed E
and E', taking account of number perturbations. $
eqn4:= -i*w*n + df(n0*vz+n*uz,z)$      % continuity eqn
n:= rhs first solve(eqn4,n)$
Pause;
Jx:= q*(n0*vx + n*ux)$
Jy:= q*(n0*vy + n*uy)$      % defining the currents
Jz:= q*(n0*vz + n*uz)$
Comment at this stage, we should be able to construct the
dielectric tensor for the homogeneous model. Since this is not
particularly useful in terms of the remaining calculations
when performed by REDUCE, it is omitted.$
Comment go straight to the Maxwell equations curlE = -dB/dt,
curlB = mu0*(J + ep0*dE/dt):        $
let n0*q^2/(ep0*m) = wp^2$      % plasma frequency
Pause;
xcpt:= df(Ex,z,2) + i*w*mu0*(Jx - i*w*ep0*Ex)$
ycpt:= df(Ey,z,2) + i*w*mu0*(Jy - i*w*ep0*Ey)$
zcpt:= i*w*mu0*(Jz - i*w*ep0*Ez)$
% zcpt defines Ez in terms of all the others.
Ez:= rhs first solve(zcpt,Ez)$
Comment now we are left with two equations involving only Ex
and Ey $
Factor Df(Ex,z,2), Df(Ex,z), Ex, Df(Ey,z,2), Df(Ey,z), Ey$
```

```
Comment for further useful manipulation of these equations,
load stage3$
end;
```

$E1_x$ and $E1_y$. During the course of the calculation, the two characteristic frequencies

$$\Omega = qB_0/m \qquad \text{(cyclotron frequency)}$$

$$\omega_p = (n_0q^2/\epsilon_0m)^{1/2} \qquad \text{(plasma frequency)}$$

appear in the working, and appropriate substitutions have been declared for them in the code. The presence of these quantities is very useful in the physical interpretation of the model structure.

B.5.4 Solving the problem

So far we have used only the model equations to generate two coupled ODEs which govern the propagation of waves in our model plasma. However, we really need to know the detailed structure of the waves, and this can only be achieved by solving the ODEs. Thus for further progress we must advance from merely coding the actual model equations into developing a mathematical approach to solving the equations. However, the motivation behind the solution technique is partly based on physical grounds which may obscure the computational goal of the exercise, and so no detailed justification of the solution process will be provided.

As a general rule, when faced with coupled ODEs whose coefficients contain trigonometrical functions of the independent variable, it may give extra insight if sin and cos are replaced by their exponential forms in the usual way:

$$\cos(x) = (e^{ix} + e^{-ix})/2 \qquad \sin(x) = (e^{ix} - e^{-ix})/2i.$$

Although this may look more complicated, especially since the imaginary number i appears, it is actually a very useful transformation when combined with the construction of two new dependent variables, viz.

$$E_+ = E_{1_x} + iE_{1_y}, \qquad E_- = E_{1_x} - iE_{1_y}.$$

The quantities E_+, E_- are important in analysing the polarisation of electromagnetic waves in a plasma. Hence the next appropriate step in solving our coupled equations is to transform the sines and cosines, and then add and subtract the two equations together in order to construct these polarisation variables in place of Ex and Ey.

This is done in the first part of STAGE3. On completion of this step, the resulting form of the equations reveals that a further simplification presents itself, namely the absorption of the common exponential factors in each equation through the further change of dependent variables from EPLUS and EMIN to

$$F_+ = E_+ e^{-i\phi'z} \qquad F_- = E_- e^{i\phi'z}.$$

The effect of all these transformations is to construct a pair of coupled ODEs with constant coefficients. This is the best possible result of the analysis, since such equations can be solved in a very simple manner.

However there is a crucial point which must be stressed at this stage. The manipulations performed by STAGE3 are not obvious, nor can they be determined only from the context of constructing computer algebra code. In fact they represent the experience of the mathematical modeller, and reflect an understanding of the physical (or mathematical) meaning of the model itself. The route to a solution is very often dictated by knowing the significance of the result. Clearly, computer algebra cannot provide this insight, but instead furnishes the modeller with the capacity to try out many ways of solving the problem without the penalty of the accompanying algebraic manipulation. The final product is a refined algebra code which achieves the desired goal.

```
%              Program STAGE3
Comment in this part of the rmf suite of codes, the sines and
cosines are converted into complex exponentials, and then E+
and E- are formed consistent with the notation of the paper.
Subsequently F+ and F- are constructed, yielding coupled ODEs
with constant coefficients.$
```

```
Pause;
ss:= xcpt + i*ycpt$    dd:= xcpt - i*ycpt$
Comment now get rid of products of trigonometrics by
substituting in favour of cos 2phi and sin 2phi:      $
let cosp^2 = (1+cos2p)/2, sinp^2 = (1-cos2p)/2,
            cosp*sinp = sin2p/2;
Pause;
Comment next we form the E+ and E- variables in the usual way$
Depend eplus, z;    Depend emin,z;
Factor df(eplus,z,2), df(eplus,z), eplus, df(emin,z,2),
       df(emin,z), emin;
let ex = (eplus + emin)/2, ey = -i*(eplus - emin)/2;
% express trigs in terms of complex exponentials. . .
let cos2p= (e^(i*2*dphi*z)+e^(-i*2*dphi*z))/2,
    sin2p = -i*(e^(i*2*dphi*z) - e^(-i*2*dphi*z))/2;
On div;
Comment next form EPLUS = FPLUS e**(i*phi*z), EMIN = FMIN
e**(-i*phi*z) in order to eliminate the z dependence in the
coefficients.  $
Pause;

Depend fplus, z;    Depend fmin, z;
Factor df(fplus,z,2), df(fplus,z), fplus, df(fmin,z,2),
       df(fmin,z), fmin;
let eplus = fplus*e^(i*dphi*z), emin = fmin*e^(-i*dphi*z);
On div;
Comment the equations should now be ready for solution! move
to stage4 for the final manipulation into a dispersion
relation.    $
end;
```

There remains the simple matter of deriving the dispersion relation itself. When an ODE has constant coefficients, such as

$$d^2y/dt^2 + \Lambda y = 0$$

then the general solutions are of the form $y = Ae^{kx}$ where $k = \sqrt{\Lambda}$. This is derived by substituting this form of solution into the differential equation, and solving the resulting algebraic relation for the unknown (k in this case). A similar procedure is used to find the nature of the general solutions to the coupled ODEs in the plasma model. Thus we substitute

$$F_+ = Ae^{ikz} \qquad F_- = Be^{ikz}$$

solve the first equation for B and thus eliminate it from the second. The resulting algebraic expression is then a fourth order polynomial in k; a dispersion relation. This polynomial then contains all the information rquired to characterise the wave solutions permitted in this medium.

Our problem is now completely solved. However, it is useful to rewrite the dispersion relation using a more compact and physically meaningful notation.

B.6 PATTERN SEARCHING

In the program STAGE4, the first half generates the polynomial in k, and the remaining code tidies up the output. In particular, we search the expression for common factors using the gcd operator, which tests for the greatest common divisor of its arguments. Once these common factors are divided out, the expression can be further simplified by adopting a condensed notation, developed partly from uniform plasma dispersion relations. In our context, we wish to use the following definitions, Stix (1962)

$$P = 1 - \sum_s \frac{\omega_{ps}}{\omega} \qquad R = 1 - \sum_s \frac{\omega_{ps}}{\omega^2} \frac{\omega}{\omega + \epsilon_s \Omega_s}$$

$$L = 1 - \sum_s \frac{\omega_{ps}}{\omega^2} \frac{\omega}{\omega - \epsilon_s \Omega_s} \qquad S = (R + L)/2$$

where the summation is over all the species of plasma particle carrying a charge of $\epsilon_s e$. In particular, when only considering one kind of particle, this reduces to

$$P = 1 - \frac{\omega_p^2}{\omega^2} \qquad RL/S = 1 + \frac{\omega_p^2(\omega_p^2 - \omega^2)}{\omega^2(\omega^2 - \Omega^2 - \omega_p^2)} \tag{B.8}$$

The task is then to make the REDUCE code search the answer for each of these patterns and make the appropriate substitutions. For the casual user of computer algebra, this is usually the most frustrating aspect of using the system. Rarely are the patterns replaced in precisely the required manner, if indeed they are identified at all. Very often, a sequence of substitutions intended as simplifying steps tend to undo previous stages, and the inexperienced user then abandons the whole process, copying down the raw output onto paper and manipulating it by hand!

The golden rule in pattern searching and substitution is to avoid ambiguity in the forms declared as substitution candidates. For example, supposing our expression contains the quantities p and q and we wish to declare a substitution for their sum and difference:

$$p + q = s \qquad p - q = d. \tag{B.9}$$

Mathematically, this is unambiguous and intuitively obvious. However, using (B.9) to construct a substitution in REDUCE will not work, since the two statements conflict: p is first declared to be $s - q$, and then subsequently $d + q$. The REDUCE code will use only the final substitution, and the candidate expression will not contain the variable s. This is because REDUCE only focuses on the first character string on the left hand side of a substitution declaration (generally speaking), and rewrites this quantity in terms of the remaining variables in the substitution declaration.

Clearly, an unambiguous method of implementing each substitution must be found. In this example, declaring

$$p = (s + d)/2 \qquad q = (s - d)/2$$

will yield the desired result. In fact, this technique has already been used in the substitutions for E_+, E_- in STAGE2.

Returning to the dispersion relation, we wish to spot the occurrences of P and RL/S as given by (B.8). Thus we must find an unambiguous way of representing the candidate patterns. One such way is the following:

$$\epsilon_0\mu_0(\omega^2 - \omega_p^2) = (\omega^2/c^2)P$$

and

$$\epsilon_0\mu_0\omega_p^2\Omega^2 = (\omega^2/c^2)\omega^2(\omega^2 - \Omega^2 - \omega_p^2)(P - RL/S)$$

This is one way of declaring the substitutions unambiguously (there are other ways, but this seems the most straightforward), and is implemented in the last half of STAGE4 (without the factor ω/c). The other substitutions involving λ_i are declared in a similar way, although the opportunity to redefine the expression is given after each substitution is declared. This is a useful tactic, in that it avoids undoing earlier forms of the expression as new ways of representing the terms are presented to the REDUCE environment. In almost every case, the definition of a useful substitution is entirely subjective. STAGE4 is merely an attempt to label gradient terms in a systematic way, and is not concerned with relations between them.

This completes the computer algebra code.

```
%               Program STAGE4
Comment here we sort out the dispersion relation itself, by
substituting for FPLUS and FMIN, and calculating the secular
determinant.$
s:= num(ss*e^(-i*dphi*z))$     dd:= num(dd*e^(i*dphi*z))$
let FPLUS = AA*e^(i*kk*z), FMIN = BB*e^(i*kk*z);
Pause;
ll:= solve(dd/e^(i*kk*z),bb);
disprin:= num(sub(first ll, ss/(AA*e^(i*kk*z))));
term:= gcd(coeffn(disprin,kk,4), coeffn(disprin,kk,2));
dr:= disprin/term;
Comment this is an attempt to match the patterns in all terms
of the coupled odes for the rmf model. Note that to avoid back
substitution, the pauses are included so that the candidates
can be redefined as new variables after each pattern
matching.$
Pause;
Let ep0*mu0*(w^2-wp^2) = PP;
Pause;
Let ep0*mu0*wp*2*omc^2/(omc^2+wp^2-w^2) = RLS - PP;
Pause;
Let ep0*mu0*u0^2*wp^2 =  (w^2-wp^2-omc^2)*rlam1;
Pause;
Let ep0*mu0*u0*omc*wp^2 = (w^2-wp^2-omc^2)*rlam2;
Pause;
```

```
Let ep0*mu0*u0*omc*dphi = (w^2-wp^2-omc^2)*rlam3;
% ---------> this is the end ......
end;
```

B.7 CONCLUDING REMARKS

In the previous sections, a physical problem was defined in mathematical terms, and used as a template for constructing a REDUCE code. The problem was then solved completely by computer algebra, in a set of four short programs. These programs represented the distilled wisdom and experience gained from a process of continuous refinement of the solution method, the details of which were not reported here.

This is possibly the most important function of computer algebra in such a context. The mathematical modeller is able to sustain an analytic attack on a particular problem in a series of repeatable, verifiable and error-free steps, each building on the last, until a complete solution method is constructed in computer algebra. This piece of code can then be used to investigate all aspects of the problem, such as explicit parameter variations, which would normally require a completely separate attack. Of course, the pattern searching and final solution method may vary with the form of the parametrisation selected, but the underlying equations cannot change, and so the modeller may proceed confidently, knowing that at least the differential equations are correct. In most cases, it is the actual generation of the equations which is the main stumbling block in theoretical analysis.

The physical model given in this paper is not in any sense an artificially constructed exercise designed as a vehicle for the presentation of REDUCE. The end result is significant in the study of plasmas with sheared fields, and the REDUCE code has permitted subsequent development of the problem beyond the simple generation of a dispersion relation. However, the problem has touched on many aspects of computer algebra which may be useful for the novice (or the sceptic). The representation of differential equations, their manipulation and deployment of intermediate solutions, and the technique of pattern searching and factorisation have all been tackled as the demand arose.

The programming is not sophisticated or particularly economical; the author is not an expert! Nevertheless, the object is to illustrate that the amateur can use simple and basic commands in computer algebra to achieve very powerful results.

ACKNOWLEDGMENTS

It is a pleasure to acknowledge fruitful discussions on both the physical and computational aspects of the problem with C C Sellar and E W Laing. The author is also grateful to the UKAEA Culham Laboratory for financial support.

REFERENCES

Boyd T J M and Sanderson J J 1969 *Plasma Dynamics* (London: Nelson)
Diver D A, Laing E W and Sellar C C 1990 On waves in spatially cyclic equilibria *J. Plasma Phys.* **43** 101–5
Diver D A and Laing E W 1990 Mode conversion and transmission in cyclic fields *J. Plasma Phys.* **43** 107–18
Kyoto Common Lisp (KCL) Research Institutes for Mathematical Science (RIMS) (Kyoto: Kyoto University)
REDUCE Users Manual Version 3.3 1987 High Level Highware Ltd, (Oxford, UK)
Stix T H 1962 *The Theory of Plasma Waves* (New York: McGraw-Hill)

APPENDIX: SELECTED OUTPUT

In this section, output is quoted from particular stages in the running of the codes. The examples are chosen as concise illustrations of the analysis at key points in the progress towards a solution.

For example, to show that STAGE1 really has solved for the velocities in terms of the electric fields, the response to the command

```
VSZ := (Q*(DF(EX,Z)*COSP*UO + DF(EY,Z)*SINP*UO + COSP*EY*OMC -
                                               2    2
          EX*OMC*SINP + EZ*I*W)) / (M*(DPHI*OMC*UO - OMC + W ))
```

showing that the required substitutions have been made. Similarly if we ask for the expression defining the perturbed number density after the first Pause in STAGE2, the response is

```
N := (NO*Q*(DF(EX,Z,2)*COSP*UO - DF(EX,Z)*DPHI*SINP*UO -
      DF(EX,Z)*OMC*SINP + DF(EY,Z,2)*SINP*UO +
      DF(EY,Z)*COSP*DPHI*UO + DF(EY,Z)*COSP*OMC + DF(EZ,Z)*I*W
      - COSP*DPHI*EX*OMC - DPHI*EY*OMC*SINP)) /
                            2    2
      (I*M*W*(DPHI*OMC*UO - OMC  + W ))
```

Obviously the status of any variable can be checked at all stages. However some of the expressions are quite lengthy, and since they are only steps on the way to the solution, they are not necessarily of interest to the reader, except where they give insight into the code structure. The next two examples show the status of the z components of the perturbed current and electric field at the end of STAGE2:

```
           2
JZ := (EPO*WP *(DF(EX,Z)*COSP*UO + DF(EY,Z)*SINP*UO +
       COSP*EY*OMC - EX*OMC*SINP + EZ*I*W)) /
                        2    2
           (DPHI*OMC*UO - OMC + W )
                      2                          2
EZ:= - (DF(EX,Z)*COSP*I*UO*WP  - EX*I*OMC*SINP*WP +
                   2                       2
       DF(EY,Z)*I*SINP*UO*WP  + EY*COSP*I*OMC*WP ) /
                        2    2    2
           (W*(DPHI*OMC*UO - OMC + W - WP ))
```

Finally it is important to state the dispersion relation as calculated by REDUCE, since this is the desired result. In fact, the output shows the dispersion relation when common factors have been divided out:

$$DISP := KK^4 *(RLAM1 + RLAM3 + 1) + KK^2 *(-2*DPHI^2 *RLAM1 -$$
$$2*DPHI^2 *RLAM3 - 2*DPHI^2 + 2*DPHI*RLAM2 - PPS*RLAM1 -$$
$$2*PPS*RLAM3 - PPS - RRS) + DPHI^4 *RLAM1 + DPHI^4 *RLAM3 +$$
$$DPHI^4 + 2*DPHI^3 *RLAM2 - DPHI^2 *PPS*RLAM1 - 2*DPHI^2 *PPS*RLAM3$$
$$- DPHI^2 *PPS - DPHI^2 *RRS + 2*DPHI*PPS*RLAM2 + PPS^2 *RLAM3$$
$$+ PPS*RRS$$

Note that the wave number is denoted by KK. This result has been used in subsequent work in plasma physics; Diver and Laing (1990).

References

Abramowitz A and Stegun J A 1964 *Handbook of Mathematical Functions* (New York: Dover)

Aizerman M A 1963 *Theory of Automatic Control* (Oxford: Pergamon)

Atadan A S and Huseyin K 1984 *J. Sound Vibration* **95** 525–30

Barenblatt G I 1987 *Dimensional Analysis* (New York: Gordon and Breach)

Boyd J B 1989 *Chebyshev and Fourier Spectral Methods* (New York: Springer)

Bravo Yuste S 1991 *J. Sound Vibration* **145** 381–90

Ceschia M and Zecchin G 1981 *IEEE Trans. Circuits Systems* **28** 456–9

Chang Y F Tabor M and Weiss J 1982 *J. Math. Phys.* **23** 531–8

Cohen H I and Fitch J P 1991 *J. Symbolic Comput.* **11** 291–305

Davenport J H, Dewar M C and Richardson M G 1992 *Symbolic and Numerical Computation for Artificial Intelligence* ed B R Donald, D Kapur and J L Mundy (London: Academic) pp 347–62

Davenport J H, Siret Y and Tournier E 1988 *Computer Algebra* (London: Academic)

Diver D A 1991 *J. Symbolic Comput.* **11** 275–89

Duncan A and Roskies R 1986 *J. Symbolic Comput.* **2** 201–6

Eastwood J W 1991 *Comput. Phys. Commun.* **64** 121–2

Fitch J P 1985 *J. Symbolic Comput.* **1** 211–27

Garrad A D and Quarton D C 1986 *J. Sound Vibration* **109** 65–78

Gates B 1987 *GENTRAN User's Manual REDUCE Version* (Santa Monica: RAND Corporation)

Genesio R and Tesi A 1992 *Automatica* **28** 531–48

Harper D 1989 *Comput. Phys. Commun.* **54** 295–305

Harper D, Wooff C and Hodgkinson D 1991 *A Guide to Computer Algebra Systems* (New York: Wiley)

Henon M and Heiles A 1964 *Astron. J.* **69** 73–9

Jordan T F 1986 *Quantum Mechanics in Simple Matrix Form* (New York: Wiley)

Lorenz E A 1963 *J. Atmos. Sci.* **20** 130–41

MacDonald N 1989 *Biological Delay Systems: Linear Stability Theory* (Cambridge: Cambridge University Press)

MacDonald N 1993 *J. Phys. A: Math. Gen.* **26** 6367–77

Mathews P M and Lakshmanan M 1974 *Q. Appl. Math.* **32** 215–8

Mickens R E 1981 *Nonlinear Oscillations* (Cambridge: Cambridge University Press)

Mickens R E 1984a *J. Sound Vibration* **94** 456–60

Mickens R E 1984b *J. Sound Vibration* **96** 277–9

Mickens R E 1986 *J. Sound Vibration* **111** 515–8

Mickens R E 1989 *Circuits Systems Signal Proc.* **8** 187–205

Pankhurst R C 1964 *Dimensional Analysis and Scale Factors* (London: Institute of Physics)

Prendergast K H 1982 *Lecture Notes in Mathematics* Vol 925 (Berlin: Springer)

Sage M L J 1988 *J. Symbolic Comput.* **5** 377–84

Schwarz F J 1985 *J. Symbolic Comput.* **1** 229–33

Scott P R 1968 *Proc. IEEE* **56** 2182–3

Seimenis J 1989 *Phys. Lett.* **139A** 151–5

Steeb W H 1993 *Int. J. Math. Phys.* C **4** 841–6

Walker S S and Connelly J A 1983 *Circuits Systems Signal Proc.* **2** 213–38

Index

T - #0234 - 101024 - C0 - 216/138/9 [11] - CB - 9780750302777 - Gloss Lamination